Scats
and
Tracks
of

Alaska

SCATS AND TRACKS SERIES

Scats
and
Tracks
of

Alaska
Including the Yukon and British Columbia

A Field Guide to the Signs of Sixty-nine Wildlife Species

James C. Halfpenny, PhD

Illustrated by Todd Telander

FALCONGUIDE®

GUILFORD, CONNECTICUT
HELENA, MONTANA
AN IMPRINT OF THE GLOBE PEQUOT PRESS

A FALCON GUIDE®

Copyright © 2007 Morris Book Publishing, LLC

Falcon and FalconGuide are registered trademarks of Morris Book Publishing, LLC.

Maps created by James C. Halfpenny, PhD © Morris Book Publishing, LLC

Library of Congress Cataloging-in-Publication Data
Halfpenny, James C.
    Scats and tracks of Alaska, including the Yukon and British Columbia : a field guide to the signs of of sixty-nine wildlife species / James C. Halfpenny ; illustrated by Todd Telander. -- 1st ed.
        p. cm. — (A FalconGuide) (Scats and tracks series)
    Includes bibliographical references (p. 140) and index.
    ISBN-13: 978-0-7627-4230-1
    ISBN-10: 0-7627-4230-5
    1. Animal tracks--Alaska. I. Title.
    QL768.H352 2006
    596.09798—dc22

                                                                2006026129

Manufactured in the United States of America
First Edition/First Printing

To buy books in quantity for corporate use
or incentives, call **(800) 962–0973**
or e-mail **premiums@GlobePequot.com.**

To Mardy and Olas Murie, consummate trackers, for educating the world about Alaska and for their tireless efforts to conserve this last frontier. Special thanks to Mardy for the years she came to my Winter Ecology classes, the great times at the Cowboy Bar (an Alaskan tradition), the marten tracks in her kitchen, and her support in following the tracks of animals. Mardy, to her dying day, fought the battle to conserve wilderness and Alaska. Without her, the trail now looks bleak.

To Diann, my alpha partner, ursophile, and tracking friend, for all her loving support and help.

JH

Contents

Acknowledgments

First and foremost, I wish to thank all my students for their years of questions and help, but most of all for the time we've shared tracking and studying in the field. I also wish to thank Lee Fitzhugh, Larry Marlow, Murie Museum, Teton Science School, and Terry McEneany (Yellowstone National Park) for his help with bird tracks.

Special thanks go to Greg Cooke, conservationist, for all the miles we skied in Yellowstone and for his undying commitment to Alaska Wilderness; to Tom Ellison, Jen Broom, and Sarah, who taught us and helped fight the good battle for the Spirit bear; and to John Rogers, boat captain and friend of the coastal brown bear, who among other great times brought us through the storm.

JH

Introduction

In the late 1970s the era of watchable wildlife arrived in the United States. Baby boomers wanted to turn to and experience the outdoors. Television brought wildlife closer than ever. Bird-watching thrived. Now more than ever, millions of people want to watch wild animals. Wildlife are not always easy to find and observe, though. Finding their tracks and signs is an exciting alternative to actually seeing the animals. Trackable wildlife adds another dimension to the outdoor experience. Todd and I wish to share that dimension, the joy of reading stories written in the soil and snow.

Upwards of ten books on tracking have been written in the United States during each decade of the twentieth century. Most are general, covering the United States or all of North America. In *Scats and Tracks of Alaska,* we focus on one biogeographic region, with details about the region's most common or characteristic species of mammals, birds, reptiles, and amphibians. (We have included a few rare species because of their particular interest or significance in a region. For example, what a coup it would be to document a polar bear or a wolverine.) We've intentionally limited the number of species covered in order to keep the information manageable. This guide is small, allowing you to carry it in a pack or pocket and use it frequently.

As your knowledge and interest in tracking grows, you may want to find additional information and help. Key references are listed in Selected Reading. For a more detailed investigation of tracking, I recommend my book *A Field Guide to Mammal Tracking in North America* (1986, Johnson Publishing, Boulder, Colorado), and titles by Olaus Murie, Louise R. Forrest, Mark Elbroch, and Paul Rezendes.

My organization will, in computer parlance, provide interactive access to expand your tracking background. A Naturalist's World (ANW), directed by myself, is an ecologically oriented company dedicated to providing educational programs and materials reflecting the natural history of North America. Diann Thompson and I run the daily business, teach classes, and lead programs. Our on-site classes provide hands-on experience and in-depth information about animals, their tracks, and the ecology of their environments. In addition to tracking classes, our field programs cover the behavior of bears, wolves, winter ecology, the northern lights, and alpine ecology. ANW also provides books, videos, slide shows, and computer programs for self-study and as teaching and field aids. You can check out ANW on the Internet at www.tracknature.com. Class schedules, product information, and information about ANW can be obtained from P.O. Box 989, Gardiner, MT 59030, phone (406) 848–9458, or on the Web at www.tracknature.com and www.trackinganimals.net.

Keep on tracking!

—*James C. Halfpenny*

About tracking

Tracking is for everyone—beginner and expert, young and old. The fun of nature's challenge is solving the mystery written in the trail. Prepare yourself by learning the background and basics of tracking before exercising your skills in the field.

Field notes and preserving tracks

To the natural history detective, the track and trail are things of great beauty and significance. They tell part of the story of an animal's life. Tracks and trails deserve to be preserved, both to increase your knowledge and as a record you can share with others. Preservation is commonly made in the form of written notes, casts, or photographs.

Perhaps the most important item in the naturalist's tool kit is the field notebook. Field notes can jog the memory and facilitate better retention of knowledge. The notes can be analyzed later and can be preserved as records of chance encounters. Writing good field notes is an art form and a science in itself. Field notes are a source of pride when shown to others and may gain recognition for recording rare and unusual events. And need I mention how quickly memories, especially for details, fade when not preserved?

While great and complex systems have been designed for complete and accurate records, there are really but three requirements for the tracker: ruler, paper, and pen. With these, every trail becomes a record for later analysis and sharing. We cannot emphasize enough the importance of enhancing your tracking experience by keeping notes to which you can later refer!

A simple 3-by-5-inch notebook and a 6-inch ruler are adequate to get started. Use a pencil or a pen that won't run if your notes get wet. To facilitate taking notes, A Naturalist's World produces a waterproof notebook that contains information about footprint groups, gaits, and how to track; data sheets for recording information; and English and metric rulers imprinted on the back cover. See the Introduction for contact information for A Naturalist's World.

Tracks may also be preserved by photographing and making casts. Good photographs can be made by any modern camera that can take a good close-up. When taking pictures, try to fill the viewfinder with the footprint. Get as close as possible. Always include a ruler or some other object in the photo to provide a sense of scale. Avoid using hats, gloves, hands, or objects without a straight edge; round edges do not lend themselves to making accurate measurements from a photo. To avoid distortion, take the photograph from directly above the track, shooting straight down. Also, step back and take photographs of the trail to show the footprints that were photographed close-up. Fast films (ASA of 200 or higher) are generally best because tracks are often found in dark places, especially ground surfaces. Modern digital cameras take excellent photographs that can easily be e-mailed to others.

Plaster casts are the old standby for preserving tracks. We suggest a casting kit that includes a one-gallon (four-liter), wide-mouthed plastic jar with a screw lid for carrying dry plaster, a narrow spatula, a plastic mixing cup such as those sold in gas station convenience stores for medium-size drinks, paper for wrapping and transporting the finished cast, and a plastic sack for cleanup. A bottle of water may be needed if water is not available on-site. Two pounds of plaster will make at least four coyote-size track casts.

Purchase plaster from a lumberyard or hardware store, as prices will be more reasonable than at a drugstore or hobby shop. Almost any plaster will work, including plaster of paris, hydrocal, ultracal, or hydrostone. Avoid getting plaster for wallboard or patching compound, however. These plasters are formulated to be slightly flexible on walls and do not get hard enough for casts. Also avoid PolyPlasters as they do not set well when the temperature is cooler than 50 degrees Fahrenheit.

Two factors are critical to preventing casts from breaking: thickness and density. In the field, thickness is assured by building a wall around the track to contain the plaster. Natural objects such as twigs, stones, and dirt may be used to make a retaining wall 0.25 to 0.5 inch (0.6 to 1.3 cm) above the track. Alternatively, walls in the form of plastic strips cut from milk cartons or other plastic containers may be taken to the field. Proper density is assured by mixing two parts of plaster to one part of water by volume (read instructions on plaster container) to create a mixture similar in consistency to thick pancake batter or a milk shake.

Place your spatula close to the track and pour onto the spatula to break the fall of the plaster into the footprint. Working quickly, so the plaster does not set and become too thick, gently pour the plaster first into the fine detailed areas of the footprint and then the rest of the print. Finally, pour the plaster to an appropriate depth (inside the retaining wall) to keep the cast from breaking. Vibrating the spatula up and down across the top of the plaster will cause it to settle evenly and create a smooth back for the cast.

Allow the plaster to dry for thirty minutes, or as long as is recommended on the plaster package. Gently pick the plaster up by digging your fingers under opposite sides of the cast, and turn the cast over onto one hand. Now wash off the dirt by rubbing the cast with your fingertips under the flowing water of a stream or hose. Do not wash the cast in a sink as plaster may clog the drain. Let the cast continue to cure for several days in a warm, dry environment. If you need to transport it, wrap the cast in paper. Never wrap the cast in plastic as trapped moisture may cause it to crumble.

While special techniques are needed for dust and snow, this procedure will allow casting in many situations. Remember, carry a plastic garbage bag and always clean up your mess. No sign of your plaster should remain to reduce the experience of others who happen by later.

Scats and bird pellets

Scats and bird pellets (also called cough pellets or castings) are often helpful for identifying an animal or completing the story written in the trail. Scats and pellets help identify not only what the animal was eating, but who the animal was. However, it should be noted that scats and pellets won't help you identify an animal with as much certainty as tracks will. Many animals make similar scats and pellets that are difficult to tell apart.

The scats of many carnivores are very similar, especially when the diet is mostly meat. Size alone does not provide a definitive answer because of the wide range of diameters produced within a species and even by a single member of a species. For example, fox produce scats ranging in size from 0.3 to 0.8 inch (0.8 to 2.0 cm), coyotes produce scats from 0.5 to 1.3 inches (1.3 to 3.3 cm), and wolves produce scats from 0.5 to 1.5 inches (1.3 to

3.8 cm), and we all know how our own scat varies in size and shape. When judging size, consider both the total quantity of scat and the size of individual pieces. Moist food produces slimmer scats, while fibrous diets produce wider scats.

Given these cautions, scat shapes can be used to identify general groups of animals (see page xix). Spherical shapes flattened top to bottom are deposited by members of the rabbit order. Elongate spheres are deposited by rodents and shrews, and at larger sizes by deer and their relatives. Long, thick cords are deposited by dogs, bears, and raccoons. Dog scats typically have tapered ends, while those from bears and raccoons are blunt. Cats also produce thick cords with blunt ends, but they tend to be constricted or even broken into short segments. Cords that loop back on themselves are produced by members of the weasel family.

Birds, in general, produce long, thin cords or shapeless, semiliquid excretions. Reptiles and amphibians may produce small elongate spheres or long, thin cords. White, nitrogenous urine deposits, found only on the scats of birds, reptiles, and amphibians, separate them from mammal scats. Scats may be confused with cough pellets. Many bird groups, including owls, raptors, crows, ravens, jays, magpies, gulls, herons, storks, flycatchers, and kingfishers, produce cough pellets in addition to scats. Birds pass digestive juices through what they have eaten to remove the nutrients. Hair, bones, beaks, claws, and other nondigestible parts accumulate in the gizzard (anterior portion of stomach), are compressed, and are coughed up as pellets. Bone and hair remnants in the pellet are easy to identify and tell much about the bird's feeding habits and even the habitats it frequents.

Pellets are grayish in color and are spherical or long and tapered at both ends. When fresh, they are covered by mucus and appear dark black. Pellets are found mainly at roosting sites and nests, and occasionally at feeding areas. They are deposited singly, but many may accumulate beneath a tree where a bird is roosting, nesting, or perching. Nitrogenous scat deposits on the ground or twigs may help verify an object as a pellet. The diameter of the throat determines the maximum diameter of the pellet. In general, large birds produce larger pellets. Shape and diameter allow one to distinguish to some degree among species. Birds generally produce two pellets per day and regurgitate just before taking flight. The time of day when feeding occurred may affect the sample of food items. For example, owls tend to feed on mammals that come out only at night, while hawks feed on animals that are out during daylight hours.

cough pellet

Anatomy and footprint nomenclature

The feet of mammals, birds, reptiles, and amphibians are anatomically complex, and that complexity shows in their footprints. Knowing something of the anatomy of their feet will aid in footprint identification and interpreting trails.

The toes of all animals are numbered from the inside of the foot out (the inside of the foot being the side closest to the animal). Therefore, in humans and other mammals, the thumb or big toe (if present) is number 1 and the little finger or little toe is number 5. In birds, toe 1 (if present) points backward.

Over evolutionary time, toes of animals have become reduced in size or have disappeared altogether. In cats and

Shapes of scats

Spheres

Rabbits and their relatives

rabbit

Spheres, elongate

Rodents, shrews, deer, and their relatives

woodrat

shrew

elk

Cords, long and thick

Wolves, bears, raccoons, and their relatives

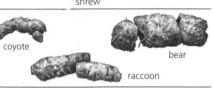

coyote

bear

raccoon

Cords, thick and often constricted

Mountain lions and their relatives

mountain lion

Cords, often folded

Weasels and their relatives

mink

Cords, long and thin, often with nitrogenous deposits

Birds, reptiles, and amphibians

Canada goose

lizard

dogs, toe 1 is absent or reduced to a small toe called a dewclaw. In deer, elk, sheep, and similar mammals, toe 1 is absent and toes 2 and 5 are reduced and form dewclaws. Toes 3 and 4, the clouts, form the cloven hoof. In pronghorn antelope, toes 1, 2, and 5 are absent. In birds, toe 5 is absent and toe 1 is often reduced and occasionally is absent. In the amphibians covered here, toe 1 has been lost from the front foot.

Track measurements

To more accurately determine the animal's foot size from its tracks, mountain lion researchers Fjelline and Mansfield (1989) developed the minimum outline method of measuring tracks.

Place your hand on a hard surface, a table for instance. Note the contact area of your hand with that surface. If your hand went no deeper into that surface, your handprint would have only one size: the minimum outline. If your hand were to sink deeper into the surface—as it would if the surface were, say, mud—it would create a series of variable outlines, each larger than the one before, as the mud flowed around the curved surface of your hand. All footprints have a minimum outline, but only prints that sink into a surface have variable outlines.

Note that while the variable outline of a footprint may only be several millimeters wider than the minimum outline, those few millimeters have a large visual effect. The human eye sees area, and area increases with the square of a linear measurement. In short, a few millimeters of width add a lot of area to a footprint.

The minimum outline size does not change for different surfaces and therefore provides a standard for comparison among surfaces. And though one animal may leave many sizes of footprints depending on surface, slope, and speed, there is only one minimum outline for every footprint that animal might leave. The minimum outline measurement is the only constant and consistent size in tracking.

To measure the minimum outline, study the bottom of a print. The *break point* where the rounded pad turns upward is the edge of the minimum outline. Use this edge to measure tracks.

Assigning the break point is a subjective judgment and no two people will always mark it at exactly the same

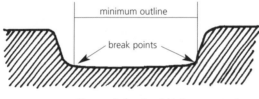

cross section of footprint in ground

point. However, testing has shown that an individual tracker using the minimum outline method can reduce personal variation in measurement and that groups of trackers using this method will also become more consistent in their measurement of tracks. Remember the computer rule GIGO: garbage in, garbage out. You cannot get quality measurements from a bad track. Quality measurements are the tracker's goal, and using minimum outline methods greatly reduces overexaggeration and variance in measurement.

All measurements in this guide are minimum outline measurements.

The measurements in *Scats and Tracks of Alaska* are mostly averages gathered from years of tracking. Averages include only animals judged to be adult. However, it is important to remember the great size variation among animals. Every animal was small once in its life, and some never get big. Males are often substantially larger than females. Regional variations in mammal sizes also occur. For example, coyotes are smaller in the southwestern United States and larger in the northeastern part of the country. Their tracks vary accordingly. Therefore, a track in the field may be considerably larger or smaller than the measurements provided. Use track measurements only as a rough guideline, not as an absolute rule. Collect your own database of track measurements for your area.

Gaits and trails

Coordinated muscle movements result in the various gaits used by animals. In the simplest form, when moving on two legs (bipedal movement), an organism can *walk, run,* and *hop.* When moving on four legs (quadrupedal movement), an organism can *walk, trot, lope, gallop, bound,* and *pronk* (also called *stot*). Though other gaits exist, we will confine our discussion to these basic gaits. Each gait leaves a characteristic pattern that may be modified by changes in speed and body angle. The combination of footprints is called the *trail.* The bipedal walk and run and the quadrupedal walk and trot result in gaits that are *symmetrical.* The right side

walk   trot

of the trail is a mirror image of the left side. The trail patterns for these gaits are the same, alternating right-left pattern, and they differ only by the stride being longer in the run and trot than it is in the walk. In the run and trot, the straddle, the distance from the right edge of the rightmost pad (see pages xxvii and xxxi) to the left edge of the leftmost pad, also tends to be narrower than it is in the walk.

Quadrupedal movement also results in gaits that are *asymmetrical* (the right half of the trail is not always a mirror image of the left), including lope, gallop, bound, hop, and pronk. These gaits produce patterns that include all four footprints (two fronts, two hinds, two rights, and two lefts) in a group separated from the next group by a space where no footprints appear.

In *gallops,* the feet, front and rear, that move first (or *lead*) will determine whether the gallop will form a Z-shaped or C-shaped pattern. When the front and hind feet on the same side lead, the pattern takes on a Z shape, called a *transverse* gallop. A right-front lead with a left-hind lead or vice versa, results in a C-shaped pattern, called a *rotatory* gallop. Thus, there are four possible gallop patterns.

C-shaped gallop | Z-shaped gallop

large print–front feet
small print–rear feet

Bounds (also known as hops and jumps) are characterized by the synchronization of the hind feet; both strike the ground at the same time, side by side. The front feet strike the ground at a different time than the hind feet. In a full bound, the front feet are synchronized and strike the ground side by side at the same time. In a half bound, only the hind feet

full bound | half bound

are synchronized and the front feet hit the ground staggered. Animals that mostly use full bounds, also called hops, live in trees (tree squirrels and songbirds), whereas those that mostly use half bounds live on the ground (ground squirrels, rabbits, and grouse).

In a *pronk* (also called a *stot*), all four feet strike the ground at the same time, with the front feet side by side and forward of the hind feet, which are also side by side. This gait is often used by deer to gain height and increase time in the air to look around.

pronk

To increase peripheral vision, non-primate mammals have eyes placed toward the sides of their heads, not flat on their face like humans. By turning sideways, a prey species can see what is pursuing it and where it needs to go to escape. The predator, by turning sideways, can see what it is chasing and where the rest of the predator pack is.

Consequently, quadrupedal mammals have evolved to use all gaits while their body is turned to the side. These *side gaits* result when the animal's heavy head deviates from the line of travel and the body turns sideways. First, the front feet respond by moving toward the side of the trail where the head is. Then, as the head turns more, the hind feet move to the side away from the head. The greater the head movement, the greater the angle of the side gait. Common examples are the side

slow side trot

side gallop

trot and side gallop that are often used by canids. These are often called a dog trot or dog gallop.

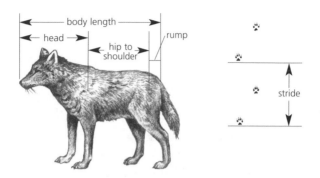

estimating mammal length from stride

An animal's size is also reflected in its gait patterns. When a mammal is walking with its normal gait, for example, the stride is 1.1 to 1.25 times longer than the distance from the hip to the shoulder joint. Using this crude relationship, body size can be judged from a walking stride. A 22-inch stride indicates a hip-to-shoulder length of 20 inches. Add to the hip-to-shoulder distance an estimate for the head length beyond the shoulder joint and an estimate of the rump length beyond the hip joint to get a total estimate of animal body length.

Speed also modifies gait patterns in trails. There are three rules governing how pattern changes as speed changes.

1. As speed increases, the hind foot lands farther forward than the front footprint on the same side. Conversely, as speed decreases, the hind foot lands farther back in relation to the front footprint.

2. As speed increases, stride increases.

3. As speed increases, straddle usually decreases.

1 x 2 x 1 lope | 2 x 2 lope | 3 x 3 lope | walk (slower) | amble (faster)

Speed changes are easily observed in quadrupedal walk and trot trail patterns. As speed increases, the hind footprint registers in front of the front print. This faster version of a walk is called an *amble*. The faster version of a trot doesn't have a name. When the animal slows to the point that the hind feet are registering behind the front prints, the animal may be stalking something. Trots are separated from walks by having a stride two or more times greater than the estimated hip-to-shoulder distance of an animal.

A slow version of the gallop is also recognizable. When a gallop slows to the point that one or more hind feet register behind the leading edge of the frontmost footprint in a group pattern, the gait is called a *lope*. The gait is still a gallop, it's just a slow gallop.

Trail measurements

The terms *stride, group, intergroup,* and *straddle* describe the size of an animal trail. The *stride* is measured from the point where a foot touches the ground surface to where the same point of the same foot next touches the surface, and consists of one group and one intergroup measurement. The *group* consists of all four footprints (two fronts, two hinds, two lefts, two rights), while the *intergroup* is the distance between groups. Gait patterns take their name from the configuration of the group. The stride provides an indication of size in a walking animal and an indication of relative speed for other gaits (see pages xxv and xxvi).

The *straddle* indicates the width of the trail and is measured from the outside rightmost pad of the outside right footprint of a group to the outside leftmost pad footprint of the same group. The outside edges of the trail are used because the inside of footprints overlap for many carnivore species. When measured from outside right to outside left, walking stride is an indicator of size.

Stride, group, and intergroup are all measured parallel to the trail, while the straddle is measured at right angles to the trail. Select a straight section of trail on level ground to measure. The slightest curve in the trail will distort the straddle measurement.

Glossary of terms

amble: a fast walk in which the hind footprint registers anterior to the front footprint. See illustration on page xxvi.

asymmetrical: not symmetrical; that is, one side is not a mirror image of the opposite side.

bound: a gait in which both hind feet strike the ground at the same time, side by side. If the front feet also land side by side, the motion is said to be a *full bound*. A *half bound* occurs when one front foot strikes the ground in front of the other. See illustration on page xxiii.

clout: term used to refer to toe 3 or toe 4 of the hoof. See illustration on page xxxii.

convergent toes: toes 2 and 4 of ducks, geese, and swans, which bend toward the foot axis, especially at the tips. Compare to *divergent* toes.

cord: See s*cat shape*.

cough pellet: remnants of bones and hair coughed up by many bird species after feeding on prey.

dewclaw: toe that over evolutionary time has become reduced in size and raised on the leg, away from the other toes; for example, toe 1 in dogs and toes 2 and 5 in deer.

diagnostic: providing certain identification of an animal or its sign.

digit: one of the toes of an animal.

digital pad: See *pad*.

digitigrade: walking on the tips of the toes. Dogs and cats, for example, are digitigrade. Tracks left by digitigrade animals rarely show a sole. Compare to *plantigrade*.

distal webbing: See *webbing*.

divergent toes: toes that are straight or turn out from the *foot axis* at the tips, specifically toes 2 and 4 of seagulls. Compare to *convergent* toes.

foot axis: imaginary line down the center of the foot. It runs between toes 3 and 4 in deer and their relatives, and down toe 3 of other mammals. In birds, the foot axis also runs down toe 3.

fringe: webbing attached to a single toe. May have a smooth edge, known as a *simple fringe*

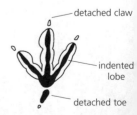

detached claw

indented lobe

detached toe

or *simple lobe,* or it may be wavy, in which case it is said to have *indented lobes.*

full bound: See *bound.*

gait: term for the type(s) of movement an animal uses when moving. Examples of gaits include *walk, amble, trot, bound,* and *gallop.* Gaits are defined by the mechanics of body movement, not by speed.

gallop: a gait in which hind feet move around the front feet and (usually) strike the ground in front of the front feet. Galloping forms distinct group patterns (two fronts, two hinds, two lefts, two rights) separated by an intergroup distance from the next set of four feet. Gallops fall into two basic patterns: Z-shaped and C-shaped.

group: a subunit of a *stride* including four footprints (two fronts and two hinds, and two lefts and two rights). The measure of the group plus the intergroup equals the measure of the stride.

gallops   bounds

half bound: See *bound.*

heel: portion of foot or track to the rear of digital and interdigital pads. In mammals, may be covered with hair, naked (without hair), or have one or more proximal pads. In reptiles and amphibians, may be textured with *tubercles.*

hop: synonymous with bound, often used in reference to *gaits* of rodents and rabbits.

indented lobe: See *fringe.*

interdigital pad: See *pad.*

length: of a track, the distance from front of toe *pads* to back of the interdigital pads, measured parallel to the *foot axis.* In mammal tracks, does not include claws. In bird tracks, does not include toe 1, but includes claws if they are attached and indistinguishable from toe pad.

line of travel: imaginary line on the ground over which the center of gravity of an animal passes.

lobe: See *fringe.*

lope: a slow *gallop,* in which at least one hind foot registers behind a front foot in a group of four footprints. See illustration on page xxvi.

mesial webbing: See *webbing*.

minimum outline: See pages xx and xxi for extended discussion.

nipple-dimple: See *scat shape*.

outer toe angle: in birds, the angle between toes 2 and 4. In perching birds less than 90 degrees and in shorebirds greater than 120 degrees.

oval: See *scat shape*.

pad: hard, calluslike structure on the sole of an animal's foot. Each toe may have a digital pad. One or more interdigital pads are located directly to the rear of the toes, and one or more proximal pads may be located directly to the rear of the interdigital pads. In deer and their relatives, there is a single pad separated from the *wall* by the *subunguinis*. In birds, a metatarsal pad may occur directly under the leg bone.

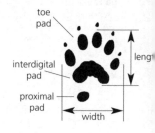

plantigrade: walking on the soles of the foot. Raccoons, bears, and humans, for example, are plantigrade. The sole of the foot usually shows in the footprint. Compare to *digitigrade*.

pronk: a *gait* in which all four feet strike the ground simultaneously and directly below the body. The *group* pattern shows two front footprints ahead of the two hind prints. Also called a *stot*. See illustration on page xxiv.

proximal pad: See *pad*.

proximal webbing: See *webbing*.

rotatory gallop: a type of *gallop* that tends to form a C-shaped *group* pattern. See illustration on page xxiii.

run: a *gait* used when moving only on two legs. It differs from a *walk* in having a longer *stride*.

scat shape: *Cords* are long pieces of scat, typically four to ten times longer than the width. Ends may be blunt or tapered. *Ovals* are pieces of scat typically two to four times longer than wide and tapered at both ends. A *nipple-dimple* shaped scat pellet has a point at one end and a depression at the other. See chart on page xix.

simple fringe, simple lobe: See *fringe*.

sole: bottom of an animal's foot. It may be covered with hair or naked, and may have one or more *pads* on it.

stot: See *pronk.*

straddle: the distance from the right edge of the rightmost pad to the leftmost edge of the leftmost pad in a *trail.* Measured at right angles to the *line of travel.*

stot

stride: the distance from the point where a foot touches the ground to the point where the same foot touches the ground again. Measured parallel to the *line of travel.* One stride is equal to a *group* plus an intergroup measurement.

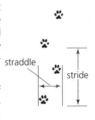

subunguinis: the soft material under the nails of humans. In deer and their relatives, refers specifically to the soft material between the *pad* and *wall.*

symmetrical: having two sides, one the mirror image of the other side.

toe pad: See *pad.*

track: refers to an individual footprint. Some measurable characteristics include *length* and *width.*

track pattern: the gross visual image of the pattern of footprints on the ground. A repeating pattern of two prints separated from the next two is called two-by and written 2 x 2. Prints may also show patterns of 3 x 3, 4 x 4, and 1 x 2 x 1. These patterns are made during a *gallop* or a *bound.* A few of these patterns are illustrated on pages xxii to xxiv.

trail: a series of footprints and associated sign that marks the passage of an animal. Some measurable characteristics include *stride* and *straddle.*

transverse gallop: a type of *gallop* that tends to form a Z-shaped group pattern. See illustration on page xxiii.

trot: a *gait* in which evenly spaced footprints alternate on right and left sides of the *line of travel.* Hind footprint registers on top of front. As speed increases, hind moves forward of front. Same patterns as a *walk,* but longer *stride.* May be done with body turned to side. See illustration on page xxii.

walk   **trot**

tubercle: rough pinhead-size protuberance on the sole of the foot of a reptile or amphibian.

unguinis: hard material forming nails in humans, hoof walls in deer and their relatives, and claws in other mammals. Composed of hair pasted together by body glues.

walk: a *gait* where evenly spaced footprints alternate on right and left sides of the *line of travel*. Hind footprint registers on top of front. As speed increases, hind moves forward of front. See illustrations on pages xxii and xxvi.

wall: hard material around the edge of each clout of a hoof. Technically the *unguinis,* which also forms human nails and animal claws.

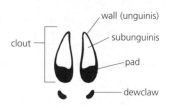

webbing: thin membrane stretched between toes of animals. The webbing may be near the tips of the toes (*distal*), about midway to the toe tips (*mesial*), or attached at the base (*proximal*). A membrane attached to only one toe is called a *fringe*.

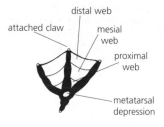

width: of a *track,* the greatest distance from the right side of the pads of a foot to the left side, whether the greatest distance is across the toes or palm pads. Measured perpendicular to the *foot axis.* In bird tracks, includes claws if they are attached and indistinguishable from toe pad.

How to use *Scats and Tracks*

Scats and Tracks of Alaska is designed for easy use in the field. The gray bars found on the edges of the pages of the track accounts will help you measure scat diameter and footprint size; each of these bars is keyed to the average size of the sign in question. A ruler is provided on the back cover. Below, we provide the background knowledge that every tracker should be familiar with before going to the field or using this book. Please take some time to study this material.

Illustrations

Illustrator Todd Telander applied his great ability to our collections of plaster casts, photographs, and slides, drawing on our experience to produce the most up-to-date and accurate illustrations possible. These drawings—made from the best specimens in a collection of thousands— represent the culmination of decades of tracking experience and are far more accurate than the tracker usually finds in tracking books. The tracks you find on the ground may not have as much detail or be as clear, but it is better to have an excellent drawing to compare to an imperfect track than to have to compare a track to a drawing lacking critical details.

How to use the track accounts

The track accounts in this guide have been grouped by similar footprint characteristics. Each track account represents a single species or a group of species with similar track characteristics. Each account is presented across a two-page spread and is conveniently divided into sections as discussed below. A brief listing of visual characteristics used to identify an animal begins each track account, appearing beneath the common and scientific names of the species. These descriptions are general, and great variability of pattern can exist among animals in the field. We recommend consulting appropriate field identification guides.

Track: A concise description of key points of footprints, to be used for identification. The accompanying track illustrations are not at actual size, but, unless otherwise noted, actual (average) length of the footprint is shown as a bar on the right side of the right-hand page. Average width is shown as a bar on the bottom of the right-hand page. In the field, place the appropriate measurement bar next to the track to compare size. To take a numerical measurement, use the ruler printed on the back cover of the book.

The tracks illustrated are all from right feet, except in the entries for birds, where both feet are pictured. Numerical measurements are given in the form *length x width*. Note that measurements of mammal tracks do not include claws, that measurements of bird tracks include claws but do not include toe 1, and that measurements generally do not include parts of the foot that often do not register in a given species's track (e.g., heels in the hind feet of some rodent species).

Trail: The average size of the stride of the most commonly used gait or gaits is given. Other common or characteristic gaits, if any, are discussed. See *track pattern* in the Glossary of Terms on page xxxi. For more information on gaits in tracking, see my *Field Guide to Mammal Tracking in North America* (1986). Gaits are displayed up the right side of the right-hand page. If the common gait is a walk or trot, however, it may not be illustrated, since all walking and trotting patterns consist of right-left alternating patterns.

Scat: A description of scat supplements the drawing. Average scat width is shown as a bar up the side of the left-hand page. In the field, place the appropriate measurement bar next to the scat to compare sizes. To take a numerical measurement, use the ruler printed on the back cover of the book. Numerical measurements of scat are given under the illustrations in the form *length x width*. In cases of small scat, only width is given; thus, a single measurement always indicates diameter.

Habitat: To aid in locating and differentiating tracks, the animal's habitat preferences are listed. Some animals with large ranges, such as the beaver, are only found in specific habitats.

Similar species: Clues are provided to help differentiate an animal's tracks from similar tracks of other species. With these clues, identification should be possible.

Other sign: Other sign of animals, besides tracks and scat, are listed or illustrated to help with identification, and simply to provide more information on animal lives.

In addition, a distribution map is provided with each account. This gives a generalized picture of where in Alaska, the Yukon, and British Columbia an animal may be found. Animals may move outside their common range. Animals that require specific habitats will of course not be evenly distributed through the shown range.

To make the best use of this guide, carry it with you into the field. When you come across an unfamiliar track or trail, open the book to the appropriate track account and place the page alongside the track for immediate on-site comparison.

Visual key to tracks

This simple key includes birds, reptiles, amphibians, and mammals. It is arranged by the number of toes that show in a good footprint, ranging from no toes to two toes to five toes. Those animals that show four toes in the front print and five toes in the hind are listed between four- and five-toed animals.

Snakes (pp. 10–11)

Series of side-to-side trail undulations.

Deer and Relatives (pp. 126–39)

Two toes form hard, cloven hoof. Dewclaws may show in deep print.

Birds with Webbed Feet (pp. 12–17, 20–21, 24–27, 40–41)

Three toes facing forward, often a fourth toe facing backward. Claws may be detached from toes. Webbing between two or more toes.

Birds without Webbed Feet (pp. 18–19, 22–23, 28–29, 42–55)

Three toes facing forward, often a fourth toe facing backward. Claws may be detached from toes.

Wolves, Dogs, and Relatives (pp. 58–65)

Four toes in front and hind prints. Claws usually present and detached. Single anterior lobe on inter-digital pad.

Cougars, Cats, and Relatives (pp. 66–71)

Four toes in front and hind prints. Claws usually absent. Double anterior lobe on interdigital pad.

Rabbits and Relatives (pp. 98-101)

Four toes in front and hind footprint. An exceptionally clear print may show a fifth inner toe in the front footprint. Pads lacking, bottom of foot covered with hair. Long hopping heel in hind print.

Salamanders (pp. 2–3)

Four toes in front print, five toes in hind. Trail wide, often with a tail drag.

Frogs (pp. 6–7)

Four toes in front print, five toes in hind. Long, slender toes. Front print faces center of trail. Distal webbing in hind print.

Toads (pp. 4–5)

Four toes in front print, five toes in hind. Front print faces center of trail. Mesial webbing in hind print. Tubercles may show on front and hind prints.

Rodents (pp. 102–25)

Most have four toes in front prints and five in hind. Beaver has five toes in front print. Front toes show a 1-2-1 grouping; hind show a 1-3-1 grouping. Long hopping heel in hind print.

Turtles (pp. 8–9)

Five toes show in front tracks and four in hind tracks. Front prints toe in and hind prints may toe out. Feet are relatively broad. Claws robust and often visible. Sometimes the claws are the only visible signs on hard ground.

Shrews (pp. 56–57)

Five slender toes present on front and hind feet. In clear prints, four interdigital and two proximal pads may be seen.

Raccoons (pp. 80–81)

Five toes in front and hind prints. Toes most often round or bulbous at ends. May have long, slender toes.

Weasels and Relatives (pp. 82–97)

Five toes in front and hind prints, though the little toe (on inside of foot) may not show. Toes in a 1-3-1 grouping. Interdigital pad is chevron-shaped. Plantigrade hind foot.

Bears (pp. 72–79)

Five toes in front and hind prints, though the little toe (on inside of foot) may not show. Toes evenly spaced. Plantigrade hind foot.

Scats
and
Tracks
of

Alaska
Including the Yukon and
British Columbia

Long-toed Salamander
Ambystoma macrodactylum

Vienna sauage–size
salamander, up to
3.0 inches (7.5 cm)
long. Moist,
smooth skin. Body
dark brown to black
green to yellowish stripe
on back.

Track: Four toes on front foot (often only three show) and five toes on hind foot. The outline of the foot may not show, just toe prints.

Trail: Walking stride is about 2.0 inches (5.0 cm). Trail has a wide straddle relative to stride and may show oscillating belly and tail drag marks.

Scat: Soft, pea-size black masses with some hint of oval shape.

scat
0.1 in
0.3 cm

egg mass

SCAT WIDTH

Habitat: Wide variety of habitats from plains to alpine tundra and rock shores of high lakes. Rotting wood and rocks near quiet waters around lakes, ponds, and streams.

Similar species: Differs from lizards by wider, oscillating tail drag and by having only four toes on front foot. Larger tracks and trail than newts.

Other sign: Eggs in egg masses are attached individually to underwater plant stems.

front
0.3 x 0.2 in
0.8 x 0.5 cm

hind
0.4 x 0.3 in
1.0 x 0.8 cm

slow walk

FRONT TRACK LENGTH

FRONT TRACK WIDTH

Boreal Toad
Bufo boreas

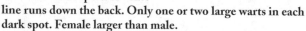

Baseball-size toad, up
to 3.5 inches (9.0 cm) long.
Body is plain brown to
gray to reddish, with warts
from yellow to red in dark
brown or black spots. A white
line runs down the back. Only one or two large warts in each
dark spot. Female larger than male.

Track: Four toes on front foot and five
on hind. Front feet face in. Two tubercles on heel of front foot. Three hind
toes face in, one forward, and one out.
Two tubercles may show on the heel of
the hind foot, and can be confused
with toes. Webbing, found only on
hind feet, extends at most halfway out
to toe tips.

Trail: Hopping or full bound stride generally 5.0 to 7.0 inches
(13.0 to 18.0 cm). Walking stride is about 2.5 inches (6.3 cm).

scat
0.9 x 0.2 in
2.3 x 0.5 cm

egg mass

toad imprint in mud

SCAT WIDTH

Scat: Dark brown to black. Long cord, up to five times longer than wide. Sometimes contains insect parts.

Habitat: Lakes, ponds, beaver ponds, and wet mixed coniferous forest.

Similar species: Differs from frog by presence of tubercles on front heels. Hind foot is narrower than frog. Walks and uses short hops more than the usually long-hopping frog.

Other sign: Long strings of egg masses on bottom of water source and floating among vegetation.

walk

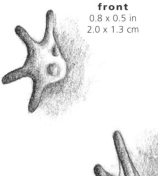

front
0.8 x 0.5 in
2.0 x 1.3 cm

hop

mesial web

hind
1.0 x 0.9 in
2.5 x 2.3 cm

FRONT TRACK LENGTH

FRONT TRACK WIDTH

Bullfrog
Rana catesbeiana

Softball-size frog, 4.0 to 8.0 inches (10.0 to 20.0 cm) long. Brownish-green to green, becoming light green on head. Legs banded with dark brown to green; small spots on back. Fold of skin around eye and large exposed eardrum.

Track: Four toes on front foot and five on hind. Front feet face in. Three hind toes face in, remaining two face forward or out. Webbing, found only on hind feet, is distal, extending most of the way out to the toe tips. In a clear track, a male's toe 2 (thumb) on front foot appears thicker at base.

Trail: Hopping stride is 24.0 inches (60.0 cm). May easily hop 72.0 inches (180.0 cm).

scat
1.5 x 0.4 in
3.8 x 1.0 cm

Scat: Brown to black, with slightly tapered ends.

Habitat: Permanent and (usually) quiet water with dense growth of aquatic plants, especially cattails, in plains, woodlands, and forest.

Similar species: Differs from toads by more webbing between toes, lack of palm tubercles on front feet, and by hopping more and at longer distances. Differs from leopard frog and chorus frog by larger feet.

Other sign: Deposits a globular egg mass.

front
1.5 x 1.0 in
3.8 x 2.5 cm

walk

distal web

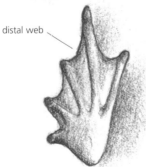

hind
1.8 x 0.9 in
4.5 x 2.3 cm

hop

FRONT TRACK LENGTH

FRONT TRACK WIDTH

Painted Turtle
Chrysemys picta

A small turtle with
smooth, unkeeled
shells and patterns of
red, yellow, and black.
Length to 8.0 inches
(20.0 cm). Females larger
than males.

Track: Five toes with claws show in front
and back prints. Claws longer on front
feet and those of the male's front feet are
two to three times longer than the
female's. Outer toe on hind foot lacks
claw. Distal webbing on front and hind
prints.

Trail: Walking stride averages 4.5 inches
(11.5 cm) with a straddle of 4.0 inches
(10.0 cm). Front feet toe in and hind feet
toe out. Hind footprint registers slightly behind. Tail drag often
visible.

Scat: Usually black semiliquid, slightly elongated clumps.

scat
0.9 x 0.2 in
2.3 x 0.5 cm

SCAT WIDTH

Habitat: Ponds, lakes, marshes, swamps, streams, and ditches with abundant vegetation.

Similar species: Differs from snapping turtles by its narrow straight trail.

Other sign: Nests are usually dug in sand in open areas where sun warms eggs.

front
0.8 x 0.7 in without claws
? 0 x 1.8 cm

hind
1.1 x 1.0 in without claws
2.8 x 2.5 cm

FRONT TRACK LENGTH

FRONT TRACK WIDTH

Snakes
various species

A variety of
snakes, from garter
snakes (*Thamnophis sirtalis*)
to milk snakes (*Lampropeltis
triangulum*) to rattlesnakes
(*Crotalus viridis*).

Rattlesnake
Crotalus viridis

Track: No footprint to describe.

Trail: Varies from 1.0 to 4.0 inches (2.5 to 10.0 cm) wide. Characterized by side-to-side undulations of the trail. The period, the distance from one curve to the next, varies by species, age, and speed of the snake. Surface material is usually pushed up at the outside of each curve. Gait is either a side-to-side undulation or sidewinding.

scat
4.0 x 0.4 in
10.0 x 1.0 cm

shed skin

SCAT WIDTH

Scat: Black or brown cord, with constrictions and undulations. White nitrogenous material often attached.

Habitat: Varies widely—from water's edge to rock outcrops to dry sand dunes.

Similar species: Resembles no other track.

Other sign: Shed skin.

**lateral
undulatory**

trail

Common Loon
Gavia immer

Large, long-bodied aquatic bird, average length 24.0 inches (60.0 cm). Dark-colored bird with black-and-white checkered back and neck band. Dark greenish head.

Track: Three toes showing, toes 2 to 4 pointing forward. Toe 1 does not register. Toes 2, 3, and 4 have claws and distal webbing.

Trail: Walking stride is 10.0 inches (25.0 cm) with a 7.0-inch (18.0-cm) straddle. Toes turn inward. Foot drag marks usually evident because the loon's legs are placed far back on its body making walking difficult and causing the loon to drag its feet.

Scat: Not known.

nest

Habitat: Freshwater lakes and rivers; near shore on ocean.

Similar species: Differs from all other aquatic birds by heavy drag marks in trail and relatively long, narrow feet.

Other sign: Large nest on reeds and brush at edge of water. Eggs appear large in small nest.

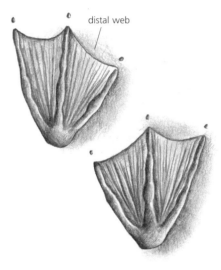

distal web

feet
5.25 x 4.3 in
13.1 x 10.8 cm

walk

TRACK LENGTH

TRACK WIDTH (50%)

White Pelican
Pelecanus erythrorhynchos

Large aquatic bird, average length more than 60.0 inches (150.0 cm), with a wingspan of more than 8.0 feet (2.4 m). White with black primary wing feathers. Large bill is yellow to orange.

Track: Four long, slender toes. Toe 1 offset to side of track. Feet *totipalmate,* with webbing between all four toes. Webbing distal and slightly convex between toes. Claws attached.

Trail: Walking stride averages 16.0 inches (40.0 cm). Toes turn inward.

Scat: Shapeless, brownish-white mass.

Habitat: Lakes, marshes, and bays. During summer, found in freshwater lakes; in winter, found in salt water.

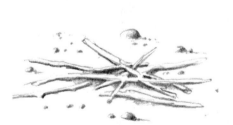

ground nest

Similar species: Differs from all other web-footed birds except cormorant by being toti-palmate. Differs from cormorant by having attached claws.

Other sign: Nests on ground in large island colonies.

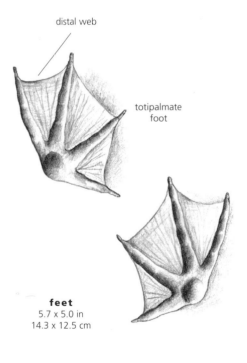

distal web

totipalmate
foot

feet
5.7 x 5.0 in
14.3 x 12.5 cm

walk

Double-crested Cormorant
Phalacrocorax auritus

Large aquatic bird, average length more than 32.0 inches (80.0 cm), with a wingspan of more than 4.3 feet (1.3 cm). Dark-colored body with orange throat patch. Crest of two white feathers behind eye, which may be difficult to see.

Track: Four long, slender toes. Toe 1 offset to side of track. Toe 4 is longest. Feet *totipalmate,* i.e., with webbing between all four toes. Webbing distal and slightly convex between toes. Claws detached.

Trail: Walking stride is about 10.0 inches (25.0 cm). Walks awkwardly on land, with a short stride for its size. Toes turn inward.

ground nest

Scat: Shapeless, nearly liquid white mass.

Habitat: Saltwater islands, bays, and cliffs. Freshwater lakes, ponds, and swamps.

Similar species: Differs from all other web-footed birds except pelican by being totipalmate. Differs from pelican by having detached claws.

Other sign: Nests in colonies on ground or in trees. Acidic scat kills trees and ground vegetation.

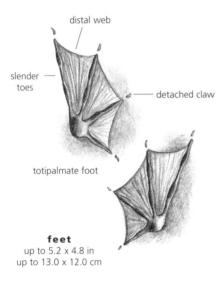

distal web

slender toes

detached claw

totipalmate foot

feet
up to 5.2 x 4.8 in
up to 13.0 x 12.0 cm

walk

TRACK LENGTH

TRACK WIDTH (50%)

Great Blue Heron
Ardea herodias

Large wading bird, average
length 45.0 inches (113.0 cm).
Male and female similar in
overall appearance: gray-blue
body, with white neck and
yellow beak. Black crown
extends on feathers off rear
of head.

Track: Four toes, toes 2 to 4 pointing forward. Small proximal web
between toes 3 and 4. Footprint is asymmetrical, with toe 1 set to inside of foot
axis (drawn through toe 3). Toe 1 is
about 1.5 inches (3.8 cm); toe 2 is longer
than 1, though shorter than 3 and 4. On
hard ground, metatarsal pad may not
show (that is, toes may appear unconnected).

Trail: Walking stride about 20.0 inches
(50.0 cm). Trail is fairly straight and feet point forward.

cough pellet

nest

rookery

Scat: Semiliquid, predominantly white. Solid cords of scat vary from 2.0 to 3.0 inches (5.0 to 7.5 cm) in length, and may contain fish, frogs, salamanders, and even small rodents. Ground beneath nests becomes coated with droppings.

Habitat: Frequents backwater eddies along riverbanks and shallow edges of lakes.

Similar species: Differs from other shore-edge tracks by large size and asymmetrical placement of toes.

Other sign: Large colonies of nests high in trees. Undigested material may be coughed up as pellets.

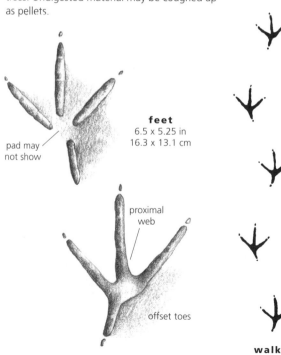

pad may not show

feet
6.5 x 5.25 in
16.3 x 13.1 cm

proximal web

offset toes

walk

TRACK LENGTH

TRACK WIDTH (50%)

Trumpeter Swan
Cygnus buccinator

Large aquatic bird, length averaging 60.0 inches (150.0 cm). Male larger than female. Adult is all white, with a black beak. Immature swan has gray plumage. Male and female identically colored.

Track: Four toes. Toes 2 to 4, which point forward, usually register. Toe 1 points rearward and only occasionally shows. Distal webbing between toes 2, 3, and 4. Toes 2 and 4 tend to converge slightly near tips. Claws are broad, blunt, and attached to toes. Feet turned in.

Trail: Walking stride is about 14.0 inches (35.0 cm).

scat
3.5 x 0.5 in
8.8 x 1.3 cm

SCAT WIDTH

Scat: Cord, five to eight times longer than wide. Often greenish and coated with white nitrogen deposits.

Habitat: Ponds, lakes, slowly flowing streams, and rivers.

Similar species: Larger than ducks and geese. Differs from pelican and cormorant by lacking webbing between toes 1 and 2. Larger than gulls, also differing from them by having convergent toes and distal webbing.

Other sign: Nests on elevated areas such as beaver lodges and muskrat houses, or on the ground. Eggs larger than eggs of Canada goose.

attached claw

feet
4.0 x 5.2 in
10.0 x 13.0 cm

distal web

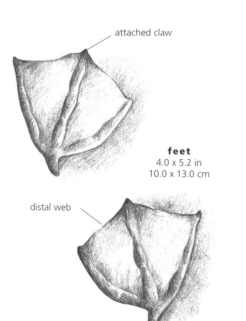

walk

FRONT TRACK LENGTH

FRONT TRACK WIDTH (50%)

Sandhill Crane
Grus canadensis

Large bird, average length 39.0 inches (98.0 cm). Appearance of males and females similar: grayish, with red crown on head, and white cheeks and chin.

Track: Four toes, toes 2 to 4 showing. Outside toes opposed by nearly 180 degrees. Toe 3 is longer than 2 and 4. Small proximal web between toes 2 and 3 rarely shows. Claws usually attached to toes, though claw of toe 1 rarely shows. Feet point forward.

Trail: Walking stride about 24.0 inches (60.0 cm). Often runs, extending its stride. Tracks have a narrow straddle, being nearly in line with one another.

Scat: Similar to, but smaller than, Canada goose. Brown in color, with some vegetation. Can contain bones of small mammals, reptiles, and amphibians.

Habitat: Meadows, marshes, grasslands, and fields.

scat
2.5 x 0.3 in
6.3 x 0.8 cm

SCAT WIDTH

Similar species: Differs from ducks, geese, swans, and herons by having only small proximal web. Differs from large raptors by lacking toe 1.

Other sign: Listen for its rattling call, which suggests to some what dinosaurs may have sounded like.

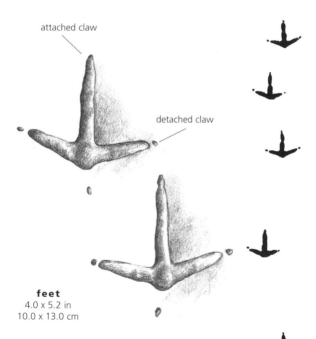

attached claw

detached claw

feet
4.0 x 5.2 in
10.0 x 13.0 cm

walk

TRACK LENGTH

TRACK WIDTH (50%)

Ducks
various species

Aquatic birds with webbed feet, varying in size from the small bufflehead through mallards (*Anas platyrhynchos*) to pintails. Length ranges from 14.0 to 24.0 inches (35.0 to 60.0 cm). Large variety in body patterns among species. Males generally more brightly colored than females.

Mallard
Anas platyrhynchos

Track: Four toes. Toes 2 to 4, which point forward, usually register. Toe 1 points rearward and may not show. Distal webbing between toes 2, 3, and 4. Webbing concave between toes. Toes 2 and 4 tend to converge near tips. Claws are broad, blunt, and attached to toes. Feet turned in.

Trail: Walking stride of a mallard (illustrated here) is about 4.0 inches (10.0 cm).

scat
2.0 x 0.25 in
5.0 x 0.6 cm

SCAT WIDTH

Scat: Pencil-size cords, four to eight times longer than wide. Often greenish and coated with white nitrogenous deposits.

Habitat: Ponds, lakes, marshes, streams, rivers, and bays.

Similar species: Tracks differ from geese and swan by smaller size. Differ from pelicans and cormorants by lacking webbing between toes 1 and 2. Differ from gulls by having convergent toes.

Other sign: Nests may be on the ground, in tree cavities, or on floating mats. Eggs roughly the size of chicken eggs, though there may be great variation among species.

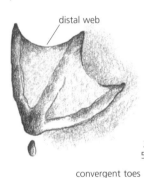

distal web

feet
2.2 x 2.4 in
5.5 x 6.0 cm

convergent toes

walk

TRACK LENGTH

TRACK WIDTH

Canada Goose
Branta canadensis

Medium-size aquatic bird, average length 30.0 inches (75.0 cm). Considerable size variation among sub-species. Black head and neck, with a white chin band. Back is olive brown. Male and female similarly colored.

Track: Four toes. Toes 2 to 4, which point forward, usually register. Toe 1 points rearward and only occasionally shows. Distal webbing between toes 2, 3, and 4. Toes 2 and 4 tend to converge slightly near tips. Claws are broad, blunt, and usually attached to toes. Feet turn in.

Trail: Walking stride is about 12.0 inches (30.0 cm).

scat
3.0 x 0.4 in
7.5 x 1.0 cm

SCAT WIDTH

Scat: Cord, five to eight times longer than wide. As long as 3.5 inches (8.8 cm). Often greenish and coated with white nitrogenous deposits.

Habitat: Ponds, lakes, marshes, streams, rivers, and bays.

Similar species: Larger than most ducks and smaller than swans. Differs from pelican and cormorant by lacking webbing between toes 1 and 2. Larger than gulls, and differing from them by having convergent toes and distal webbing.

Other sign: Nests on ground, sometimes on cliff ledges, and in abandoned heron and raptor nests. Eggs larger than chicken eggs.

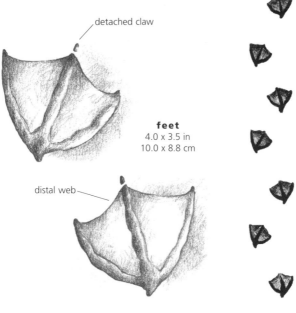

detached claw

feet
4.0 x 3.5 in
10.0 x 8.8 cm

distal web

walk

TRACK LENGTH

TRACK WIDTH

Hawks
various species

Medium-size birds, up to 25.0 inches (63.0 cm) in length, with wingspans of 50.0 inches (125.0 cm). The red-tailed hawk (*Buteo jamaicensis*) is a highly variable, dark-colored hawk with red tail feathers. Light phase has a dark belly band.

Red-tailed hawk
Buteo jamaicensis

Track: Four wide, robust toes. Toes 2 to 4 point forward. Claws are long, sharp, and not attached to the toe print.

Trail: Walking stride of a red-tailed hawk (illustrated here) is about 12.0 inches (30.0 cm). Hops or runs after prey on the ground. Claws may drag in soft mud, creating a fringe along sides of footprint.

Scat: Semiliquid, primarily white with some brown intermixed. Unlike owls, scat is ejected with force, sometimes leaving a trail. Whitish piles and vertical streaks below nests.

cough pellets

Habitat: Woodland and open country with scattered trees.

Similar species: Smaller than eagles. Differ from owls by having three toes pointing forward. Differ from geese and swans by lacking webbing. Differ from herons and cranes by having symmetrical feet.

Other sign: Cough pellets may be 1.5 to 4.0 inches (3.8 to 10.0 cm) long.

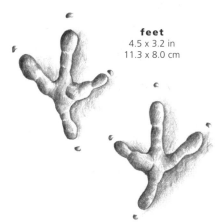

feet
4.5 x 3.2 in
11.3 x 8.0 cm

walk

TRACK LENGTH

TRACK WIDTH

Eagles
two species

Large birds, averaging 35.0 inches (90.0 cm) in length, with wingspans of 80.0 inches (200.0 cm). Brown bodies. Adult golden eagle (*Aquila chrysaetos*) has golden feathers over head and

Bald eagle
Haliaeetus leucocephalus

neck. Adult bald eagle (*Haliaeetus leuco-cephalus*) has white head, neck, and tail feathers.

Track: Four wide, robust toes. Toes 2 to 4 point forward. Lacks webbing and metatarsal pad. Claws are long, sharp, and not attached to the toe print.

Trail: Walking stride about 18.0 inches (45.0 cm). Golden eagle will run after prey on the ground.

Scat: Semiliquid, primarily white with some brown intermixed.

Habitat: Golden eagle found in hilly areas and hunts over open country. Bald eagle usually found near lakes and rivers.

cough pellet

urine stain on rock

Similar species: Larger than hawk's track, which is less than 3.0 inches (7.5 cm). Differ from owls by having three toes pointing forward. Differ from geese and swans by lacking webbing. Differ from herons and cranes by having symmetrical feet.

Other sign: Cough pellets may be 5.0 x 1.5 inches (12.5 x 3.8 cm). Nests may be 6.0 feet (2.0 m) in diameter.

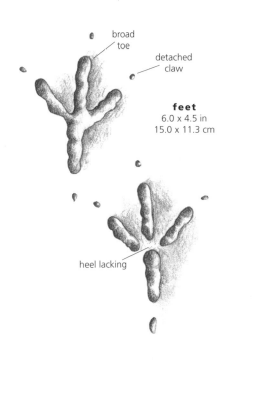

broad toe

detached claw

feet
6.0 x 4.5 in
15.0 x 11.3 cm

heel lacking

walk

TRACK WIDTH (50%)

Spruce Grouse
Dendragapus canadensis

Size comparable to a small chicken, 17.0 inches (43.0 cm) in length. A dark brown to gray-black bird with white spots on chest, especially the male. Orange tail feather tips with red combs above eyes.

Track: Four toes, with toes 2 to 4 pointing forward. Toe 1, relatively short, may not show. Toes are relatively wide and lack webbing. Claws detached. In winter, a fringe of scales makes toes wider.

Trail: Walking stride is about 9.0 inches (23.0 cm). Trail is straight and feet toe in. In deep snow, there can be a 4-inch (10.0 cm) wide trough.

Look for wing marks in the snow as well as body impression under brush and evergreen trees.

hard form

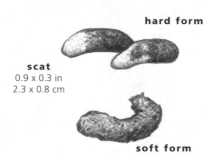

scat
0.9 x 0.3 in
2.3 x 0.8 cm

soft form

Scat: Light to dark brown, sometimes with white nitrogenous covering. Content includes buds, berries, and sawdust.

Habitat: Spruce forests with moss ground cover.

Similar species: Tracks larger than bobwhite. Differ from other forest birds by wide, robust toe size and short toe 1. Grouse trails can be differentiated from those of other forest birds by their short stride.

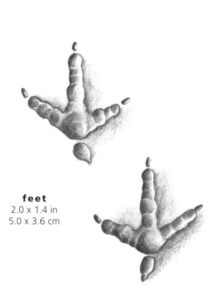

feet
2.0 x 1.4 in
5.0 x 3.6 cm

walk

TRACK LENGTH

TRACK WIDTH

Ptarmigan
Lagopus species

At an average length of
14.0 inches (35.0 cm),
larger than the Ameri-
can robin. Male and
female similarly col-
ored: brown- or black-
and-white-mottled
body in summer, with white on wings. All white in winter.
Red eye combs. Three species, rock (*L. mutus*), willow
(*L. Lagopus*), and white-tailed (*L. leucurus*), live in the
region. They cannot be told apart by their tracks.

Track: Four toes, toes 2 to 4 pointing
forward. Toe 1, relatively short, may not
show. Toes are relatively wide and lack
webbing. In winter the heavily feathered
foot shows broad, diffuse toe imprints,
and detail may be masked. Feet point
forward to slightly inward.

Trail: Walking stride is about 6.0 inches
(15.0 cm).

scat
1.25 x 0.25 in
3.1 x 0.6 cm

SCAT WIDTH

**snow nest and wing
marks from takeoff**

Scat: Light to dark brown, sometimes with white nitrogenous covering. Includes buds, berries, and sawdust. About 1.5 inches (3.8 cm) long. In winter, deposited in snow nest in large mass of fifty or so scats.

Habitat: Rocky alpine slopes, meadows, and willow patches of high mountains.

Similar species: Differs from other forest birds by wide, robust toes and short toe 1. Trail shows short strides and wide straddle. Feathers on feet differentiate from other forest birds. Toe angle is narrower than grouse.

Other sign: Flies into or burrows under the snow to roost.

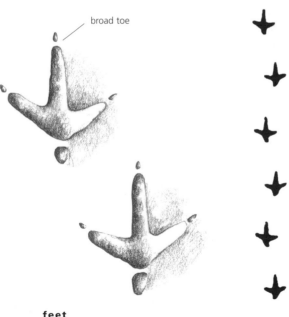

broad toe

feet
2.4 x 2.0 in
6.0 x 5.0 cm

walk

TRACK LENGTH

TRACK WIDTH

Coot
Fulica americana

Medium-size aquatic bird, average length 15.0 inches (38.0 cm). Slate-black body, with white beak extending into small brown forehead shield.

Track: Four toes showing, toes 2 to 4 pointing forward. Toe 1 angles inward. Toes 2, 3, and 4 have fringe of webbing with indented lobes. Long, pointed claws, especially those on toe 1, may be separated from toes.

Trail: Walking stride 10.0 inches (25.0 cm); tends to wander when walking. Foot axis parallel to line of travel.

Scat: White liquid.

Habitat: Freshwater lakes and ponds having shallow water where reeds and rushes grow.

Similar species: Differs from all other aquatic birds by the indented lobes on each toe.

Other sign: Floating nest built from cattails, sedges, and rushes, rising several inches above the water.

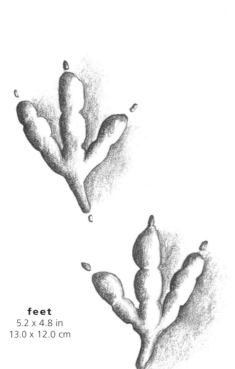

feet
5.2 x 4.8 in
13.0 x 12.0 cm

walk

TRACK LENGTH

TRACK WIDTH (50%)

Shorebirds
various species

Many birds, such as sandpipers, killdeer, curlews, and snipes. Average length varies from 6.0 to 18.0 inches (15.0 to 45.0 cm). All have similar footprints and differentiation is difficult.

Spotted Sandpiper
Actitis macularia

Track: Four narrow toes, although toe 1 may not show. Toes 2 to 4 face forward and are nearly symmetrical around toe 3. Outer toe angle often greater than 120 degrees. Small, proximal webbing between toes 2, 3, and 4 may be visible, though curlews and sandpipers have proximal webbing only between toes 3 and 4.

Trail: Shorebirds are constantly running along the water's edge. Stride varies from 4.0 to 20.0 inches (10.0 to 50.0 cm).

beach with beak holes

Scat: Small and semiliquid. Browns, green and white mixed.

Habitat: Water's edge at lakes, rivers, streams, wastewater treatment plants.

Similar species: Differ from song- or perching birds by the weak showing of toe 1, which in perching birds is strong and used to grasp branches. Outer toe angle of songbirds is less than 90 degrees. Shorebirds walk, but most songbirds hop.

Other sign: Myriad roundish holes where beak pushed into the sand in pursuit of insects.

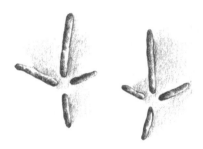

feet
1.25 x 1.0 in
3.1 x 2.5 cm

walk

Gulls
various species

Average length 25.0 inches (63.0 cm). Pale-gray back, white head. Tips of primary feathers black. Yellow bill with red spot; pink legs. Herring gull (*Larus argentatus*) illustrated here. Many

Herring gull
Larus argentatus

species, such as Bonaparte's, Franklin's, Ring-billed, and Herring gulls. Length varies from 11.0 to 30.0 inches. Gregarious species of open beaches of lakes, oceans, and rivers.

Track: Four toes. Toes 2 to 4 (forward-pointing) show. Toe 1 may register only slightly or not show at all. Webbing relatively straight between toes. Toes 2 and 3 tend to diverge, especially at the tips.

Trail: Walking stride of herring gull is about 13.0 inches (33.0 cm). Feet turn slightly inward.

Scat: Semiliquid. Primarily white, with indistinguishable contents.

cough pellet

Habitat: Along coast and on inland lakes and rivers. Nest in colonies on ground or cliffs, usually on islands. Nest is made of grass or seaweed. A scavenger, the herring gull is also found at dumps.

Similar species: Differ from ducks, swans, and geese by having divergent toes. Smaller than swans and geese. Differ from coot by having webbing between toes.

Other sign: Cough pellets containing bones, fish scales, urchin parts, and garbage. Shell fragments from dropping mussel shells onto rocks from high in the air.

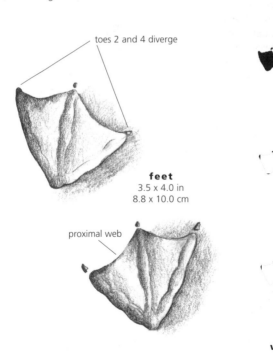

toes 2 and 4 diverge

feet
3.5 x 4.0 in
8.8 x 10.0 cm

proximal web

walk

TRACK LENGTH

TRACK WIDTH (50%)

Owls
various species

Considerable variation in length, from the saw-whet owl (*Aegolius acadicus*), 8.0 inches (20.0 cm), to the short-eared owl (*Asio flammeus*) illustrated here, 15.0 inches (38.0 cm), to the great horned owl (*Bubo virginianus*), 25.0 inches (63.0 cm). All species have immobile eyes offset by facial disks of feathers. Great variability in appearance between species. Typical body colors are grays, browns, and reddish browns.

Short-eared owl
Asio flammeus

Track: Four broad toes, with two paired and facing forward. Toe 4 position is not fixed and may face back or out. Lack webbing and metatarsal pads. Claws long and detached from footprint. Tracks of short-eared owl illustrated here.

Trail: Walking stride varies considerably among species, from 3.0 to 10.0 inches (7.5 to 25.0 cm).

cough pellet

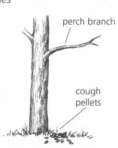

perch branch

cough pellets

perch with pellets

Scat: Semiliquid, primarily white.

Habitat: Forested areas. Some species, such as barn owls, will readily use human structures.

Similar species: Differ from most birds in toes 2 and 3 being paired, nearly parallel, and pointing forward. Differ from woodpeckers by toes being wide and robust, and by toes 1 and 4 being much shorter than toes 2 and 3.

Other sign: Cough pellets below a roost. Diameter of cough pellets ranges from 0.25 to 1.0 inch (0.6 to 2.5 cm) and is directly related to the size of the owl. Pellets are shiny and black when new but turn gray with age.

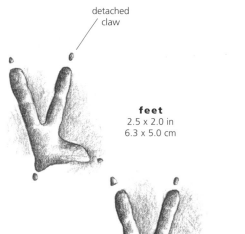

detached
/ claw

feet
2.5 x 2.0 in
6.3 x 5.0 cm

walk

TRACK LENGTH

TRACK WIDTH

Woodpeckers
various species

Considerable variation in size. The flicker (*Colaptes auratus*) is a medium-size woodpecker, slightly larger than the American robin, average length 12.0 inches (30.0 cm). Male has brown-barred back, black chest, white rump, and red or black whisker stripe, and is yellow under wings. Female lacks whisker stripe.

Flicker
Colaptes auratus

Track: Four toes, with two parallel and pointing forward. Toes 1 and 4 point backward and are not equal in length. Strong, rigid tail feathers may show on ground.

Trail: Walking stride of the flicker (illustrated here) is about 3.0 inches (7.5 cm). Hopping stride is about 4.0 inches (10.0 cm).

Scat: Cord, about four or more times longer than wide. Often contains undigested parts of insects.

Habitat: Open woodlands, dense forests, and around towns.

scat
1.0 x 0.25 in
2.5 x 0.6 cm

SCAT WIDTH

Similar species: Differs from three-toed woodpecker by presence of toe 1. Differs from other birds its size by having two toes pointing forward.

Other sign: Excavates and nests in tree cavities. Does not add bedding material to cavity nest.

tree with beak holes

toes point forward

feet
1.75 x 0.75 in
4.4 x 1.9 cm

walk

Gray Jay
Perisoreus canadensis

Small bird with gray body, blackish nape, and black beak. Average length about 11.0 inches (28.0 cm). This friendly bird, whose food-foraging habits earned it the name camp robber, will eat from your hand. Anthropologists tell us that butchered bones indicate that gray jays have been at our campfires for nearly 9,000 years.

Track: Narrow track with four narrow toes, toes 2 to 4 facing forward and not widely splayed. Toe 1 as long as toes 2, 3, and 4. Toe 1 broader than other toes. Lacks webbing and metatarsal pad. Claws long, especially on toe 1.

Trail: Hopping stride is 4.0 to 5.0 inches (10.0 to 12.5 cm). Occasionally walks.

Scat: Semiliquid, brown to black with white intermixed.

Habitat: Coniferous trees where it feeds on seeds, insects, and even carrion.

Similar species: Differs from songbirds by larger size and relatively narrow footprint. Smaller and narrower than crows and ravens.

Other sign: Nest, on horizontal branch or tree crotch, is compact and occasionally cemented with mud.

hop

walk

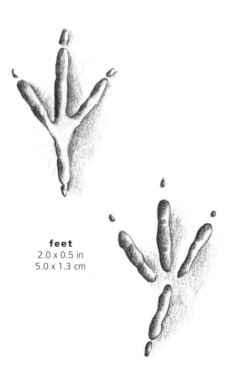

feet
2.0 x 0.5 in
5.0 x 1.3 cm

TRACK LENGTH

TRACK WIDTH

American Robin
Turdus migratorius

Perhaps the most recog-
nized North American
bird, the robin is a small
bird with a dark gray
back and orange chest
and stomach. The head
is darker with a white
ring around the eye.
Average length is 10.0
inches (25.0 cm).

Track: Four toes, toes 2 to 4 facing forward. Length of toe 1 nearly equals
toes 2 and 4. Toe 3 is longer than 1 and
bulbous at end. Outer toes spread.
Lacks webbing and metatarsal pad.
Claws short and may be detached from
footprint.

Trail: Walking stride varies considerably,
but is about 3.0 inches (7.5 cm). Also
runs, with a longer stride.

Scat: Semiliquid to pasty and white.

Habitat: Open woodland habitat to urban areas. Uses conifer or deciduous trees for nesting. Usually seen in the grass searching for earthworms.

Similar species: Differs from songbirds by wide spreading toes, creating a relatively wide, curving footprint. Smaller and narrower than jays and crows whose toes do not splay.

Other sign: Cough pellets up to 3.0 x 0.5 inches (7.5 x 1.3 cm). Caches food in forks of trees and, often, by burying.

hop

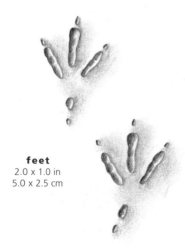

feet
2.0 x 1.0 in
5.0 x 2.5 cm

walk

TRACK LENGTH

TRACK WIDTH

Crow
Corvus brachyrhynchos

Medium-size (17.0 in / 43.0 cm)
black bird with strong
beak (smaller than
raven's). Sides of tail
are parallel in flight,
not wedge shaped. Black feet
and legs.

Track: Four toes, three facing forward. Toe 1 nearly equals toes 2, 3, and
4. Lacks webbing. Metatarsal pad present. Claws long and often detached
from footprint. The footprint length of
2.5 inches (6.3 cm) includes toe 4,
which adds 0.7 inch (1.8 cm).

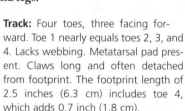

Trail: Walking stride varies but is about
5.0 inches (13.0 cm). Crows both walk
and hop and may run with a long
stride.

Scat: Semiliquid brown and white, but may contain remnants of
food from their omnivorous diet.

cough pellet

Habitat: Roadside, woodlands, farms, orchards, and lake and ocean shores.

Similar species: Raven track and trail is much larger than crow's. Lack the paired forward facing toes of owls. Lack long toe 1 of hawks.

Other sign: Cough pellets up to 1.0 x 0.4 inches (2.5 x 1.0 cm). Pellets may contain berries, seeds, nuts, insect parts, and snails, among other items of its varied diet.

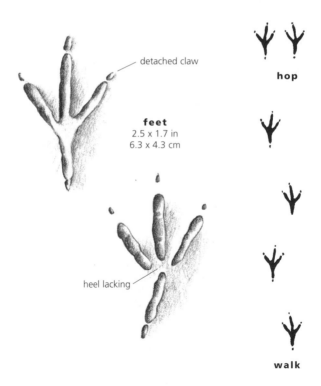

detached claw

feet
2.5 x 1.7 in
6.3 x 4.3 cm

heel lacking

hop

walk

TRACK LENGTH

TRACK WIDTH

Common Raven
Corvus corax

**Large black bird,
length averaging 24.0
inches (60.0 cm). Beak
heavy. Tail is wedge shaped
in flight. Size varies
considerably, though
male is larger than
female.**

Track: Four toes, toes 2 to 4 facing forward. Length of toe 1 nearly equals toes 2, 3, and 4. Lacks webbing and metatarsal pad. Claws long and detached from footprint.

Trail: Walking stride varies considerably, but is about 20.0 inches (50.0 cm). Also runs, with a longer stride.

Scat: Semiliquid, brown, black, and white. Often oily. May contain remnants of its omnivorous diet.

Habitat: Mountains, especially where carcasses of deer and elk can be found, and at garbage dumps. Will beg food from picnickers.

cough pellet

Similar species: Track and trail of the common crow are diminutive versions of the raven's. Lacks the paired forward facing toes of owls. Lacks the long toe 1 of hawk's. Smaller than eagle's.

Other sign: Cough pellets up to 3.0 x 0.5 inches (7.5 x 1.3 cm). Caches food in forks of trees and, often, by burying.

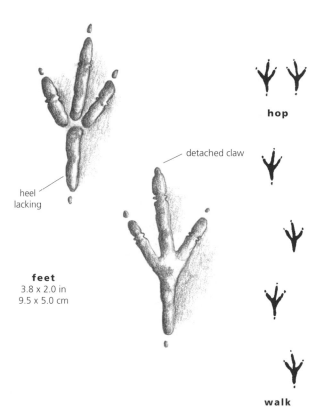

detached claw

heel lacking

feet
3.8 x 2.0 in
9.5 x 5.0 cm

hop

walk

TRACK LENGTH

TRACK WIDTH

Black-billed Magpie
Pica pica

Black and iridescent green body, wings, and unusually long tail. Average length about 20.0 inches (50.0 cm). Belly is white. Beak is black.

Track: Four medium-wide toes, three facing forward. Toe 1 nearly as long as toes 2, 3, and 4. Lacks webbing and metatarsal pad. Claws long and detached from footprint.

Trail: Walking stride is about 6.0 inches (15.0 cm).

Scat: Semiliquid, brown with white intermixed.

cough pellet

Habitat: Lower mountains, in open woodlands, thickets, along stream edges. Often found near human habitation.

Similar species: Differs from songbirds by larger size and relatively wide toes. Smaller than crows and ravens.

Other sign: Caches food in trees and under bark. Cough pellets are 1.25 x 0.4 inches (3.2 x 1.0 cm).

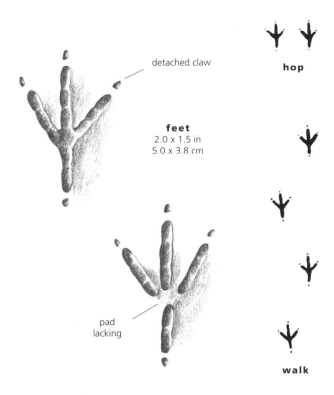

detached claw

hop

feet
2.0 x 1.5 in
5.0 x 3.8 cm

pad
lacking

walk

TRACK LENGTH

TRACK WIDTH

Shrews
various species

Masked shrew
Sorex cinereus

Smaller than mice, less than 0.25 ounce (7.0 g). Long, pointed nose. Minute eyes and ears. Color brown to black, with gray to white belly. Eat mostly insects. The masked shrew (*Sorex cinereus*) is illustrated here.

Track: Five slender toes are present on front and hind feet. In clear prints, four interdigital and two proximal pads may be seen.

Trail: Hopping stride seldom more than 2.0 inches (5.0 cm). The group of tracks is less than 1.0 inch (2.5 cm) long. Seldom is the stride more than three times the group.

Scat: Usually small pellets with tapered ends.

Habitat: Found everywhere from grasslands to alpine areas. Look for tracks in wet, fine mud of riparian areas or in snow along logs

—— tapered ends

scat
0.08 in
0.2 cm

insect remains

SCAT WIDTH

or the edges of buildings. Wood piles and leaf litter make good homes.

Similar species: Differ from mice and voles by having five toes on front foot.

Other sign: After eating, leave body parts from insects they have killed. Often burrow just below the surface of the snow, opening tunnels that partially collapse and expose their route. Trails in the snow radiate from holes like spokes of a wheel.

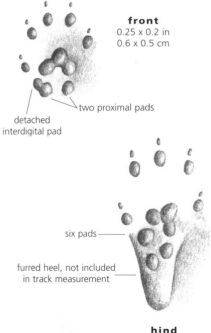

front
0.25 x 0.2 in
0.6 x 0.5 cm

two proximal pads

detached
interdigital pad

six pads

furred heel, not included
in track measurement

hind
0.3 x 0.2 in
0.8 x 0.5 cm

bound

FRONT TRACK LENGTH

FRONT TRACK WIDTH

Red Fox
Vulpes vulpes

Border collie–size, 6.0 to 15.0 pounds (3.0 to 7.0 kg). Reddish yellow, with black stockings and a white tip on the tail. Regional color phases include silver, black, cross, and bluish gray. Long, pointed ears and elongate, pointed muzzle.

Track: Claws prominent. One lobe on the leading edge of the interdigital pad. Inside toe slightly larger than outside. A ridge of callus present across interdigital pad, but difficult to detect on hind footprint. Front foot larger than hind.

Trail: Trotting stride averages 32.0 inches (80.0 cm). Typically uses a trotting gait and, occasionally, a 2 x 2 trot with body turned to the side. Walks more than coyote, especially in shrubs.

Scat: Often has tapered tail. Composition varies. Mouse or rabbit fur, berries, and insects are common. Bird feathers and plant remains often present.

scat
2.0 x 0.6 in
5.0 x 1.5 cm

log

SCAT WIDTH

Habitat: Found in a variety of habitats from brush to croplands to mixed hard- and softwood forest. Prefers edges, where hunting for small mammals is good. Also found in urban areas, where cover is available during the daytime. Not found in dense forests.

Similar species: Differs from other canids by having a ridge of callus on the interdigital pad. Track tends to be larger than that of gray fox, and usually shows claws. Differs from bobcat in having only one lobe on the interdigital pad and claws (usually) showing.

Other sign: Multiple dens are used each season. Often digs own den. A given den may be used for several years. Look for small bones around den entrance. Scat has a diagnostic musky odor, produced by a musk gland on the top of the tail. Learn to identify this unique foxy odor. Foxes tightrope-walk on narrow logs. May take over woodchuck dens.

side trot

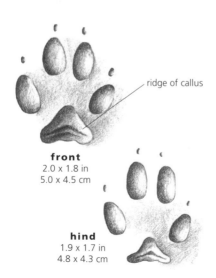

ridge of callus

front
2.0 x 1.8 in
5.0 x 4.5 cm

hind
1.9 x 1.7 in
4.8 x 4.3 cm

trot

FRONT TRACK LENGTH

FRONT TRACK WIDTH

Arctic Fox
Alopex lagopus

The size of a large domestic cat, 3.3 to 7.5 pounds (1.5 to 3.5 kg), with disproportionately small ears and legs. Two color forms, which change from summer to winter, exist: grayish brown (summer) to white (winter) and dark bluish gray or chocolate brown (summer) to bluish gray (winter). Large fluffy tail that acts like a weather vane in heavy wind, causing the fox to be blown parallel to the wind. Bottoms of feet are mostly hair-covered in winter.

Track: Small for a canid. Claws often do not show. Front foot larger than hind.

Trail: Generally lopes with a stride of 25.0 inches (64.0 cm). Hair-covered foot registers indistinctly in winter snow.

Scat: Usually has a tapered tail. Scat consists mostly of small mammals and birds, berries, and carrion.

scat
2.0 x 0.6 in
5.0 x 1.5 cm

SCAT WIDTH

Habitat: Hunts and scavenges along shore edges eating seabirds, eggs, and small mammals. Follows polar bears to scavenge off their kills.

Similar species: Smaller than coyote and red fox, whose claws show better. Lacks the ridge of callus on the interdigital pad of the red fox.

Other sign: Dens in burrows in the ground and snow. Forms or day beds on high and prominent ridges and rocks.

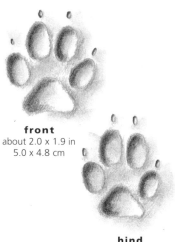

front
about 2.0 x 1.9 in
5.0 x 4.8 cm

hind
1.8 x 1.7 in
4.6 x 4.3 cm

FRONT TRACK LENGTH

FRONT TRACK WIDTH

Coyote
Canis latrans

Collie-size canid, 20.0 to 25.0 pounds (9.0 to 11.0 kg). Male larger than female. Color varies from completely gray to tan to rust. Long, pointed ears and long, narrow muzzle.

Track: Claws usually present. One lobe on the leading edge of the interdigital pad. Inside toe slightly larger than outside. Front foot larger than hind.

Trail: Trotting stride averages 41.0 inches (103.0 cm). Often uses a trot with body turned to the side, leaving a 2 x 2 track pattern. Often lopes, leaving a C-shaped pattern.

Scat: Varies from pure black animal protein to mostly hair with some bones. Tips tapered into long tails.

Habitat: An animal of the open brush country, the coyote digs its den on exposed hilltops or ridges with a view of surrounding area. Where persecuted, may den in a more secluded location.

Similar species: Even adult track is smaller than that of a wolf pup. Track may overlap in size with red fox, but lacks callus ridge

scat
3.0 x 0.6 in
7.5 x 1.5 cm

SCAT WIDTH

of red fox. Track larger than gray fox, and usually shows claws. Differs from bobcat by showing claws and by having one lobe on leading edge of interdigital pad.

Other sign: Marks territory with urine and scat piles. Scat pile locations may be used repeatedly. Uses feet to scratch near scat piles, spreading odor from scat and foot glands to identify territory.

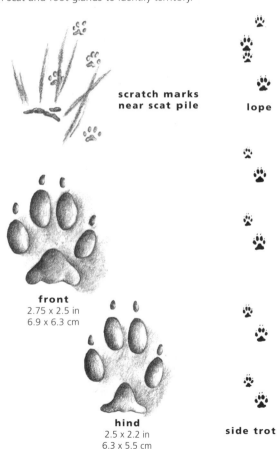

scratch marks near scat pile

lope

front
2.75 x 2.5 in
6.9 x 6.3 cm

hind
2.5 x 2.2 in
6.3 x 5.5 cm

side trot

FRONT TRACK LENGTH

FRONT TRACK WIDTH

Wolf
Canis lupus

About German shepherd size, probably less than 100.0 pounds (45.0 kg). Male larger than female. Color varies from completely black to gray. Short, rounded ears and short, wide, blocky muzzle. The wolf apparently recolonizing the northeast from Canada is a smaller member of the genus.

Track: Track about size of a softball. Claws usually present; one lobe on leading edge of interdigital pad. Inside toe slightly larger than outside. Front foot larger than hind. Given the indefinite origin of wolves, measurements below are approximate and possibly represent upper limits.

Trail: Trotting stride averages 62.0 inches (155.0 cm). Often uses a C-shaped gallop or a trot with the body turned to the side, leaving a 2 x 2 pattern.

Scat: Varies from pure black, toothpaste-like animal protein to mostly hair with some bones. Tips tapered into long tails.

scat
4.0 x 1.25 in
10.0 x 3.2 cm

SCAT WIDTH

Habitat: Found in all habitats. Tends to use cover when possible, moving in forest or at forest-meadow edge.

Similar species: Track larger than other canids. At 60 days of age, track larger than adult coyote's. Differs from mountain lion by having one lobe on the leading edge of the interdigital pad and usually showing claws. Differs from wolverine and bear by having only four toes, with the large toe on the inside.

Other sign: The alpha wolf, the dominant member of the pack, marks its territory by urinating on raised objects along the trail. Blood observed in the female's urine stain during January or February may indicate readiness to breed. Scratch marks beside urine stains or scat are territorial markings and are usually made with hind feet.

side trot

urine mark of alpha wolf

front
4.25 x 4.0 in
10.6 x 10.0 cm

hind
3.75 x 3.25 in
9.4 x 8.1 cm

gallop

FRONT TRACK LENGTH

FRONT TRACK WIDTH (50%)

Bobcat
Felis rufus

Size of a collie, weighing 13.0 to 35.0 pounds (6.0 to 16.0 kg). Males larger than females. Overall color reddish to yellowish brown, with dark spots or streaks and whitish underside. Ears have tufts at tips. Back of ears and top of tail tip black. Tail is short or bobbed, about 4.0 inches (10.0 cm) long.

Track: Front track is round or wider than long. Hind track may be longer than wide. Claw impressions are usually absent. Toes form a slight arc and toe 3 leads. The leading edge of the interdigital pad has two lobes. Inside toe distinctly larger than outside toe.

Trail: Walking stride is about 20.0 inches (50.0 cm). Usually walks, but bounds with hind feet placed side by side when chasing prey. Winter trails often show random vertical leaps, perhaps signaling that the bobcat has jumped after a flying bird.

Scat: Tends to be constricted and, if dry, separates at constrictions into segments. Ends usually blunt. Dry scat falls apart. Scat from a fresh kill may form a cord of uniform diameter.

broken constriction

scat
3.0 x 0.8 in
7.5 x 2.0 cm

SCAT WIDTH

Habitat: Prefers dense cover of swamps and forests, especially with rocky ledges. Open agricultural land is not used. Rock piles, caves, and high rocky ledges are important for bearing young.

Similar species: Differs from coyote and other canids by lacking claws, having two lobes on the leading edge of the interdigital pad, and having toe 3 leading. Substantially smaller than both lion and lynx.

Other sign: Scent marks made by urine, scat, and anal glands. Scrapes dirt or snow over urine and scat. Scratches from rubbing glands are apparent on snow. Caches food by burying.

walk

vertical leap from hind feet

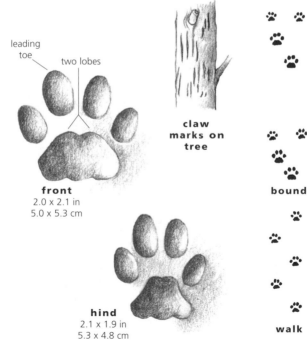

leading toe

two lobes

front
2.0 x 2.1 in
5.0 x 5.3 cm

claw marks on tree

bound

hind
2.1 x 1.9 in
5.3 x 4.8 cm

walk

FRONT TRACK LENGTH

FRONT TRACK WIDTH

Canada Lynx
Felis lynx

Larger than a collie, with male averaging 22.0 pounds (10.0 kg) and female 19.0 pounds (9.0 kg). Very long legs and big feet. Reddish to yellowish brown overall, with dark spots or streaks and whitish underside. Ears have tufts at tips. Tail tip is black on top and bottom. Tail is short or bobbed, about 4.0 inches (10.0 cm) long.

Track: Diameter of a softball and indistinct because the feet are mostly covered with hair, and because pads are reduced in size. Feet are large, for better support on snow.

Trail: Walking stride is about 28.0 inches (70.0 cm). Walking gaits are common, but lynx does trot more than bobcat. Winter trails often show random vertical leaps, perhaps signalling that the lynx has jumped after a flying bird.

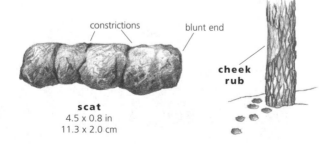

constrictions   blunt end

cheek rub

scat
4.5 x 0.8 in
11.3 x 2.0 cm

SCAT WIDTH

Scat: Tends to be constricted and, if dry, separates at constrictions into segments. Ends blunt. Dry scat falls apart. Scat from a fresh kill may form a cord of uniform diameter.

Habitat: Found in dense conifer forests interspersed with rocky ledges and downed timber, both of which are used for security and denning. Forest edges, which provide food for the lynx's major prey, snowshoe hare, are critical.

Similar species: Track differs from other felids by being inherently indistinct. Interdigital pad is relatively small when compared to bobcat and mountain lion—check closely. Differs from canids by two lobes on interdigital pad and claws not showing.

Other sign: Birth dens occur in hollow logs, stumps, and clumps of timber. Adult lynx does not cover scat. Lynx urinates (scent marks) up to twenty-five times per mile.

fast trot

vertical leap from hind feet

trot

walk

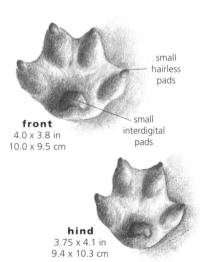

front
4.0 x 3.8 in
10.0 x 9.5 cm

small hairless pads

small interdigital pads

hind
3.75 x 4.1 in
9.4 x 10.3 cm

FRONT TRACK LENGTH

FRONT TRACK WIDTH (50%)

Mountain Lion
Puma concolor

Larger than a German shepherd, with male averaging 145.0 pounds (66.0 kg) and female about 120.0 pounds (54.0 kg). Color gray to red, often called tawny, with whitish underside. Back of ears and tail tip black to brown. Tail is more than half the length of the body. Also called cougar or puma.

Track: Track is diameter of a baseball. Front track round or wider than long, and hind track longer than wide. Claw impressions are usually absent. Toes form a slight arc and toe 3 leads. Leading edge of the interdigital pad has two lobes. Inside toe distinctly larger.

Trail: Walking stride is about 36.0 inches (90.0 cm). Usually walks, but bounds with hind feet placed side by side when chasing prey.

Scat: Scat from a fresh kill may form a cord of uniform diameter with very slight constrictions; ends usually blunt. As the carcass on which a lion is feeding dries out, the lion's scat tends to develop constrictions, eventually falling apart when diet becomes very dry.

scat
4.0 x 1.25 in
10.0 x 3.1 cm

Habitat: Habitat is that of its main prey, deer. Open woodlands with rock ledges and grass (for deer) preferred. Often found in riparian zones with trees.

Similar species: Track differs from wolf by the presence of two lobes on the leading edge of the interdigital pad, by having toe 3 leading, and by usually not showing claws. Differs from wolverine and bears by having only four toes and large toe inside.

Other sign: Often buries scat by scraping dirt over it with front feet. Scraped ground material may conceal food caches. Male will rake up basketball-size patches of brush and urinate on them to mark home range.

Comments: Breeding populations not known from area. Proof of non-feral or non-escaped animals is needed.

scraped ground around scat

bound

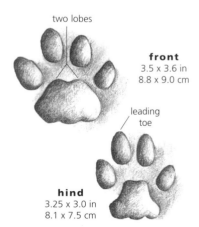

two lobes

front
3.5 x 3.6 in
8.8 x 9.0 cm

leading toe

hind
3.25 x 3.0 in
8.1 x 7.5 cm

walk

FRONT TRACK LENGTH

FRONT TRACK WIDTH

Black Bear
Ursus americanus

Calf-size bear, female averaging 120.0 pounds (54.0 kg) and male about 300.0 pounds (135.0 kg). Male grows faster and obtains larger size than female. Color varies from black to brown to blond to red. Special color phases, occasionally awarded subspecific status, include the white spirit or kermode bear (*U. a. kermode*) of British Columbia and the blue or glacier bear (*U. a. emmonsi*) of southern Alaska and British Columbia.

Track: Claws on front foot, seldom longer than toes, are usually present. Little toe is set back from rest of toes. Hind print has a large, humanlike heel. Outside toe is larger than others.

Trail: Walking stride 35.0 to 40.0 inches (88.0 to 100.0 cm). Usually ambles, a fast walk where the hind foot oversteps the front. Gait is pigeon-toed. Lopes in a C-shaped pattern or a side gallop.

Scat: Normally contains vegetation and is sweet-smelling. When

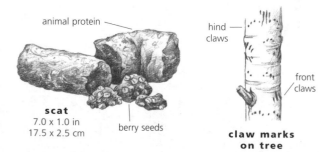

animal protein

scat
7.0 x 1.0 in
17.5 x 2.5 cm

berry seeds

hind claws

front claws

claw marks on tree

SCAT WIDTH

feeding on carcasses, scat varies from black to brown, with mostly hair and some bones. Ants often found in scat. Tips have a short taper or are blunt.

Habitat: Forest, seldom venturing far into wide openings. Thick understory vegetation and abundant food sources are critical.

Similar species: Differs from grizzly bear by smaller size, shorter claw length, and more curved arc of toes. Differs from wolverine by having toes tightly packed and having a wedge-shaped interdigital pad. Differs from lion and wolf by having five toes.

Other sign: Claws trees, rips open logs, digs into ant piles, and turns over rocks and scat as it looks for insects.

side lope

lope

amble

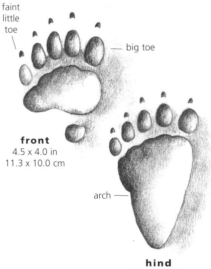

faint little toe

big toe

front
4.5 x 4.0 in
11.3 x 10.0 cm

arch

hind
7.0 x 3.5 in
17.5 x 8.8 cm

FRONT TRACK LENGTH

FRONT TRACK WIDTH (50%)

Grizzly Bear
Ursus arctos horribilis

Small, cow-size bear, female averaging about 350 pounds (160 kg) and male about 450 pounds (200 kg). Both sexes have hump over shoulders. Color varies from black to brown to blond. Light tips on individual hairs create the grizzled appearance for which grizzlies are named. All brown/grizzly bears are of the same species, *U. arctos*, but subspecific status is awarded by some authors; genetics studies suggest other differences and hybridization. For tracking, differences in footprints are apparent. The interior grizzly is a solitary bear with large personal space into which humans must tread lightly.

Track: Claws on front foot, typically more than 1.5 times longer than toes. Webbing occurs between toes; look carefully. Hind print has a large, humanlike heel. Outside toe larger than others.

Trail: Walking stride is 50 to 60 inches (125 to 150 cm). Usually ambles, a fast walk where the hind foot oversteps the

scat
7.0 x 1.5 in
17.5 x 3.8 cm

SCAT WIDTH

front. Gait is pigeon-toed. Lopes in a C-shaped pattern or uses a side gallop.

Scat: Normally contains vegetation and is sweet-smelling. When feeding on carcasses, scat varies from black to brown, with mostly hair and some bones. When feeding in alpine areas may contain only parts of moths. Ants often found in scat. Tips have a short taper or are blunt.

Habitat: High forests to alpine tundra meadows. Stays close to cover, foraging at forest-meadow edge.

Similar species: Differs from black bear by flatter arc of toes, longer claws, and webbing. Differs from wolverine by having toes tightly packed and having a wedge-shaped interdigital pad. Differs from wolf and lion by having five toes.

Other sign: Claws trees, rips open logs, digs into ant piles, digs out roots and rodent caches, and turns over rocks and scat as it looks for insects. Rubs body against tree trunks, smoothing bark and leaving hair.

side lope

lope

amble

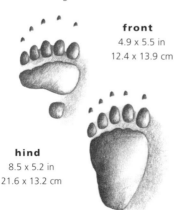

front
4.9 x 5.5 in
12.4 x 13.9 cm

hind
8.5 x 5.2 in
21.6 x 13.2 cm

FRONT TRACK LENGTH

FRONT TRACK WIDTH (50%)

Kodiak and Coastal Brown Bears
Ursus arctos middendorffi and *U. a. dalli*

Kodiak bears may be bull-size, weighing up to about 1,500.0 pounds (680.0 kg). They are the largest brown bears and are found only on Kodiak and Afognak Islands.

Coastal brown bears may be cow-size, weighing up to about 1,200.0 pounds (550.0 kg). Color usually a shade of brown. All brown/grizzly bears are of the same species, *U. arctos*, but subspecific status is award by some authors; genetics studies suggest differences and hybridization. For tracking, differences in footprints are apparent. The Katmai and coastal brown bears are social bears used to feeding in congregations at salmon streams. They have small personal space into which humans are often welcome, allowing nearby viewing but with great care.

Track: Claws on front foot, typically more than 1.5 times longer than toes and blunt. Webbing occurs between toes; look carefully. Hind print has a large, humanlike heel. Outside toe larger than others.

scat
10.0 x 2.0 in
25.0 x 5.0 cm

SCAT WIDTH

Trail: Walking stride of Kodiak is 80.0 to 90.0 inches (200.0 to 230.0 cm) while coastal bear is 60.0 to 80.0 inches (150.0 to 200.0 cm). Usually ambles, a fast walk where the hind foot oversteps the front. Gait is pigeon-toed. Lopes in a C-shaped pattern or uses a side gallop.

Scat: Normally contains vegetation and is sweet-smelling. When feeding on salmon, scat may be a semiliquid mass with fish scales and fish odor. When feeding on berries, scat may be blue, but when feeding on grasses there is green texture.

Habitat: At the mouth of coastal streams, in berry patches, and grassy meadows.

Similar species: Differs from black bear by flatter arc of toes, longer claws, and webbing. Differs from grizzly bear by larger feet and trails.

Other sign: Remains of salmon and wide trails beat down in the grass. Repeated use of trail leaves footprint holes as deep as 2 feet in the muskeg.

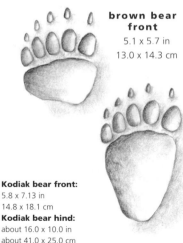

brown bear front
5.1 x 5.7 in
13.0 x 14.3 cm

Kodiak bear front:
5.8 x 7.13 in
14.8 x 18.1 cm
Kodiak bear hind:
about 16.0 x 10.0 in
about 41.0 x 25.0 cm

brown bear hind
8.7 x 5.7 in
22.0 x 14.3 cm

FRONT TRACK LENGTH (BROWN BEAR)

FRONT TRACK WIDTH (BROWN BEAR) (50%)

Polar Bear
Ursus maritimus

Large bull-size bear, weighing up to about 1,800.0 pounds (820.0 kg). On average, the polar bear is the largest bear, although individual Kodiak bears compete in size. Color is  yellowish white in summer and winter but may be stained with dirt. Feeds almost exclusively on seals.

Track: Claws on front foot, typically as long as toes but curved and sharp. Webbing occurs between toes; look carefully. Hind print has a large, humanlike heel. Outside toe larger than others.

Trail: Walking stride is 80.0 to 100.0 inches (200.0 to 250.0 cm). Usually ambles, a fast walk where the hind foot oversteps the front. Gait is pigeon-toed. Lopes in a C-shaped pattern or uses a side gallop.

scat
about 10.0 x 2.0 in
about 25.0 x 5.0 cm

SCAT WIDTH

Scat: During summer may contain remains of vegetation and berries. Most of the year contains remains of seals. Shells of blue mussels (*Mytilus edulis*) are found in fall scat when bears are waiting for the ocean pack ice to freeze.

Habitat: Coastal shores along the Arctic Ocean.

Similar species: Densely haired foot shows in footprints and separates polar bear tracks from other bears.

Other sign: Holes dug in kelp for beds or feeding. Appear to feed on the blue mussels in these kelp holes.

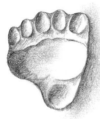

front
6.8 x 7.8 in
17.3 x 19.8 cm

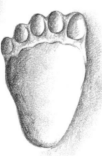

hind
about 12.0 x 9.0 in
about 31.0 x 23.0 cm

amble

FRONT TRACK LENGTH (50%)

FRONT TRACK WIDTH (50%)

Raccoon
Procyon lotor

Stocky, smaller
than a collie, with
broad head and
bushy tail. Male
averages 18.0
pounds (8.0 kg)
and female 16.0
pounds (7.0 kg). Gray to
black overall, with black rings on the tail and a black mask
on a white face.

Track: Five slender toes, slightly bulbous
on the ends. Feet resemble small human
hands and feet. Hind foot has a long,
naked heel.

Trail: Walking stride averages 27.0 inches
(68.0 cm). Roll of hips during walk causes
hind foot to register beside the opposite
front print. C-shaped gallop is common.

Scat: Highly variable, but often black, even-diameter cord with
blunt ends. Often contains crayfish or fruit. Deposited singly or in
dung heaps containing scat from perhaps several raccoons. **May
carry a parasite that is fatal to humans. Do not smell scat,
and wash hands after touching.**

Habitat: River and stream drainages are prime habitats, but storm
drains in cities may also provide refuge. Woodpiles in and around
towns.

scat
3.0 x 0.75 in
7.5 x 1.9 cm

SCAT WIDTH

Similar species: Differs from bear in having slender toes. Differs from river otter by lack of webbing. Larger than mink.

Other sign: Digs holes in streambanks to get at crayfish. Leaves piles of crayfish skeletons and claws. Digs for worms in lawns.

**sign left
while
fishing for
crayfish**

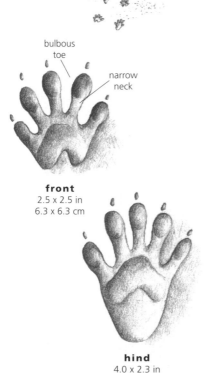

bulbous
toe

narrow
neck

front
2.5 x 2.5 in
6.3 x 6.3 cm

hind
4.0 x 2.3 in
10.0 x 5.8 cm

gallop

walk

FRONT TRACK LENGTH

FRONT TRACK WIDTH

Weasels
Mustela species

Long-tailed weasel
M. frenata

Three species, with long, slender bodies, varying in size from a regular to a foot-long hot dog. Pointed, flat skull with small ears. Males up to twice as large as females. Largest males weigh about 1.0 pound (0.5 kg). Overall color is brown, with a white belly. In winter, northern individuals turn entirely white. Hairy, slender tail. The small least weasel (*M. nivalis*) lacks the black tip on the tail found in the ermine (*M. erminea*) and the long-tailed weasel (*M. frenata*, shown here).

Track: Wide track. Five toes, in 1-3-1 grouping. Little toe, on inside of foot, often does not register. Interdigital pad chevron-shaped. Heel seldom shows. Difficult to distinguish among species.

Trail: Galloping stride varies from 8.0 to 30.0 inches (20.0 to 75.0 cm). Side-by-side tracks, when examined closely, show one track slightly in front of the other—a gallop. In snow, a drag mark may be found between front and hind prints, sometimes forming a dumbbell shape.

Scat: Long, slender cord, usually with black, toothpaste-like animal protein or hair. Cord tends to fold back on itself. Tapered at both ends.

folded back ———

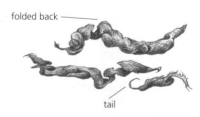

scat
1.5 x 0.1 in
3.8 x 0.3 cm

tail

SCAT WIDTH

Habitat: Prefer dense, low ground cover to open areas. Found in habitats where their prey, rodents, occur in high densities. Trails often lead from one rodent den to another. Travel in snow and ground burrows of other mammals.

Similar species: Differ from other mustelids by their smaller size and the drag mark commonly located between twin track patterns in the snow.

Other sign: Routes seldom follow a straight line, often having many sharp turns. Scat often deposited on raised objects.

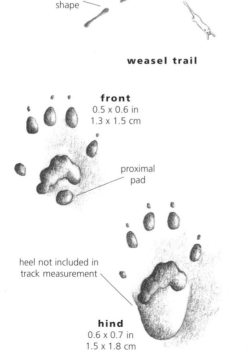

erratic route

dumbell shape

weasel trail

front
0.5 x 0.6 in
1.3 x 1.5 cm

proximal pad

heel not included in track measurement

hind
0.6 x 0.7 in
1.5 x 1.8 cm

foot drag

gallop

2 x 2 lope

FRONT TRACK LENGTH

FRONT TRACK WIDTH

Marten
Martes americana

Size of a small house cat, but slender, 1.0 to 2.0 pounds (0.5 to 1.0 kg). Male 20 percent larger than female.

Pointed, flat skull with small ears. Overall color golden brown, with orange to yellow chest patch. Edges of ears are white. Hairy, slender tail.

Track: Five toes, in 1-3-1 grouping. Little toe, on the inside of foot, sometimes does not register. Interdigital pad is a chevron. Proximal pad may show in front footprint. Heel often shows. Feet are well furred in winter, making tracks indistinct.

Trail: Gallop stride averages 22.0 inches (55.0 cm). Mostly gallops; a variety of 2 x 2, 3 x 3, and 4 x 4 patterns will be found.

Scat: Long, slender cord, tending to fold back on itself. Black or brown in color, occasionally with hair. Tapered at both ends.

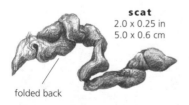

scat
2.0 x 0.25 in
5.0 x 0.6 cm

folded back

SCAT WIDTH

climbs tree, jumps out

Habitat: Old-growth forest, but adaptable to many forest habitats. Prefers mature conifer and mixed forests. Needs tall, hollow, or broken trees for denning. Found near its prey, squirrels and red-backed voles.

Similar species: Differs from weasel by its larger size. Lacks the webbed toes of the mink. Also differs from mink by use of terrestrial habitat. Smaller than fisher and occupies areas with deeper snow.

Other sign: Frequently burrows beneath snow and climbs up trees; look for tracks that end at a tree trunk. Scratch marks show where stomach was dragged over objects that protrude from the ground or snow to scent mark.

4 x 4 gallop

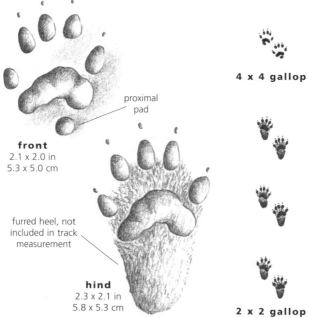

proximal pad

front
2.1 x 2.0 in
5.3 x 5.0 cm

furred heel, not included in track measurement

hind
2.3 x 2.1 in
5.8 x 5.3 cm

2 x 2 gallop

FRONT TRACK LENGTH

FRONT TRACK WIDTH

Mink
Mustela vison

**Size of a small
domestic cat, but
slender, 1.5 to 3.5
pounds (0.7 to 1.5
kg). Male 10 percent larger
than female. Pointed, flat skull with small
ears. Overall color is dark brown, with white
spots on chin and chest. Hairy, slender tail. Webbing occurs
between the toes.**

Track: Five toes, in 1-3-1 grouping. Little toe, on the inside of foot, sometimes does not register. Webbing shows between toes in tracks; look carefully. Interdigital pad chevron-shaped. Proximal pad may show in front footprint. Heel seldom shows.

Trail: Bounding stride averages 14.0 inches (35.0 cm). Bounds more than weasels, but a gallop, averaging 20.0 inches (50.0 cm), is also common.

Scat: Long, slender cord, usually tending to fold back on itself. Black or brown in color, occasionally with hair. Tapered at both ends. Often contains remains of fish or crayfish. May be oily and smell fishy. Fish oil keeps scat composed of fish scales from falling apart until oil evaporates, then scales scatter on the ground.

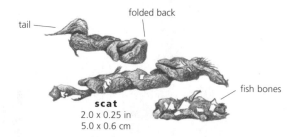

tail

folded back

fish bones

scat
2.0 x 0.25 in
5.0 x 0.6 cm

SCAT WIDTH

Habitat: River- and streambanks. Seldom far from water.

Similar species: Differs from other small mustelids by having more webbing between toes. Larger than weasels. Use of aquatic habitat is an important clue for separation from marten. Tracks and trail much smaller than those of otter.

Other sign: Mink make "post offices," repeated scat deposits on logs exposed above water's edge. Strong, musky, almost skunklike odor from anal scent glands.

fast walk

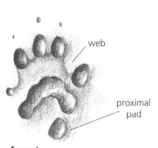

front
1.7 x 1.8 in
4.3 x 4.5 cm

hind
1.8 x 1.9 in
4.5 x 4.8 cm

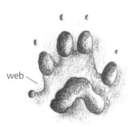

bound

FRONT TRACK LENGTH

FRONT TRACK WIDTH

Fisher
Martes pennanti

Larger than a
large domestic
cat, but slender,
7.5 to 12.0 pounds
(3.0 to 5.0 kg). Male
larger than female.
Pointed, flat skull with small ears. Color is dark brown. Long
bushy tail.

Track: Five toes, in 1-3-1 grouping. Little toe, on the inside of foot, sometimes does not register. Interdigital pad chevron-shaped. Proximal pad may show in front footprint. Heel seldom shows. Claws short. Feet are not well furred, which, in winter, makes toes appear clearly in tracks.

Trail: Galloping stride averages 28.0 inches (70.0 cm). Mostly gallops, but walking, 3 x 3, 1 x 2 x 1, and 4 x 4 gallop patterns are also common.

Scat: Only mustelid scat that frequently contains porcupine quills. Long, slender cord, usually tending to fold back on itself. Black or brown in color, occasionally with hair. Tapered at both ends.

Habitat: Old-growth forest, especially among conifers and large timber, and in swamp areas. Upland hardwood stands where porcupines den. Will use young forest stands following fire or timber

folded back

tail

scat
3.5 x 0.5 in
8.8 x 1.3 cm

SCAT WIDTH

harvest. Avoids open areas without overhead cover, but will travel on roads and trails.

Similar species: Lacks the webbed toes of the mink. Differs from mink by habitat and use of terrestrial sites. Larger than mink and marten and occupies areas of shallower snow. Smaller than wolverine and makes more frequent use of trees for walkways and nests.

Other sign: Porcupine skins turned inside out. Snow trails may reveal frequent trips up trees. Walks on logs to avoid deep snow. Drags stomach over objects that protrude from the ground or snow to scent mark, leaving scratches. Travels on packed trails of snowshoe hare.

1 x 2 x 1 lope

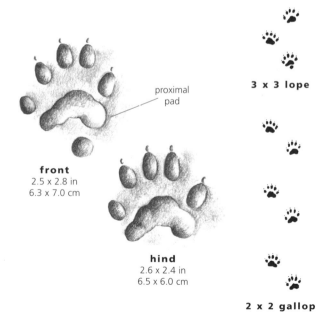

3 x 3 lope

front
2.5 x 2.8 in
6.3 x 7.0 cm

proximal pad

hind
2.6 x 2.4 in
6.5 x 6.0 cm

2 x 2 gallop

FRONT TRACK LENGTH

FRONT TRACK WIDTH

River Otter
Lutra canadensis

Body and tail form a
4-foot-long cylinder
that tapers to a hairy,
pointed tail. Weight
varies from 10.0 to 30.0
pounds (5.0 to 14.0 kg).
Male slightly larger than female. Overall
color a rich, dark brown, with silver-brown belly. Webbed
toes on front and hind feet.

Track: Large webbed foot is diagnostic, but look closely because webbing may be difficult to see. Hind foot is very wide. Five toes, in 1-3-1 grouping. Little toe, on the inside of foot, sometimes does not register. Interdigital pad chevron-shaped. Proximal pad often shows. Hairless heel on hind foot.

Trail: Walking stride averages 19.0 inches (48.0 cm). Loping stride averages 32.0 inches (80.0 cm). Loping gait patterns are usually turned to the side.

Scat: Usually contains fish remains, including scales and vertebrae. The texture is oily and the smell fishy. Fish oil keeps scat composed of fish scales from falling apart. Scat decomposes as oil evaporates, eventually falling into a pile of scales.

fish parts

scat
5.0 x 1.0 in
12.5 x 2.5 cm

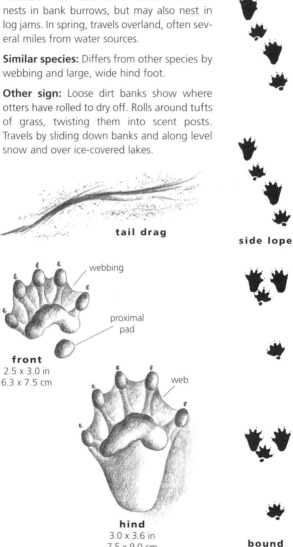

Habitat: River- and streambeds. Lives and nests in bank burrows, but may also nest in log jams. In spring, travels overland, often several miles from water sources.

Similar species: Differs from other species by webbing and large, wide hind foot.

Other sign: Loose dirt banks show where otters have rolled to dry off. Rolls around tufts of grass, twisting them into scent posts. Travels by sliding down banks and along level snow and over ice-covered lakes.

tail drag

side lope

webbing

proximal pad

front
2.5 x 3.0 in
6.3 x 7.5 cm

web

hind
3.0 x 3.6 in
7.5 x 9.0 cm

bound

FRONT TRACK LENGTH

FRONT TRACK WIDTH

Wolverine
Gulo gulo

Body like a badger's, but
heavier, up to 35.0
pound (16.0 kg). Male
larger than female.
Head broad, rounded,
and flat. Color varies
from overall dark brown
with light blond side stripes to nearly all blond. Short tail.

Track: Track diameter of a baseball. Five toes, in 1-3-1 grouping. Little toe, on the inside of foot, sometimes does not register. Interdigital pad is a chevron. Proximal pad and heel often show. Distinct winter tracks because feet are not well furred.

Trail: Gallop stride averages 35.0 inches (90.0 cm). Mostly 2 x 2 and 3 x 3 gallop patterns.

Scat: Long, medium-diameter cord occasionally folding back on itself. Black or brown, occasionally with hair. Tapered at both ends. Similar to large coyote scat.

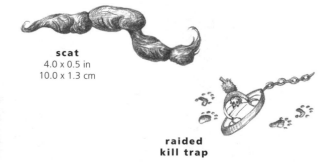

scat
4.0 x 0.5 in
10.0 x 1.3 cm

**raided
kill trap**

Habitat: Not habitat-specific. Wanders widely in all seasons, though often found where there are wintering deer, elk, or moose.

Similar species: Larger than mink, marten, and fisher, and only rarely climbs trees. Differs from wolf and lion by having five toes. Differs from bear by having 1-3-1 toe grouping and chevron-shaped interdigital pad.

Other sign: Scavenges on old carcasses, including those of animals caught in kill traps. Revisits sites to dig carcasses from under snow. Drags stomach over objects that protrude from ground or snow to scent mark, leaving scratches. Trails cross large openings in trees and are often found above tree line. Travels on packed trails and roads.

3 x 3 gallop

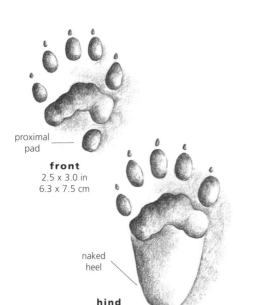

proximal pad

front
2.5 x 3.0 in
6.3 x 7.5 cm

naked heel

hind
3.0 x 3.6 in
7.5 x 9.0 cm

1 x 2 x 1 gallop

FRONT TRACK LENGTH

FRONT TRACK WIDTH

Badger
Taxidea taxus

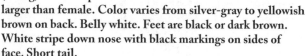

Border collie–size, with flat body, long hair, and long, shovel-like claws, about 18.0 pounds (8.0 kg). Male 25 percent larger than female. Color varies from silver-gray to yellowish brown on back. Belly white. Feet are black or dark brown. White stripe down nose with black markings on sides of face. Short tail.

Track: Diameter of a golf ball, with long front claws, nearly as long as rest of footprint. Five toes, in 1-3-1 grouping. Little toe, on the inside of foot, sometimes does not register. Interdigital pad chevron shaped. Proximal pad often shows. Front footprint larger than hind.

Trail: Walking stride averages 14.0 inches (35.0 cm). Walking is most common, but trotting, with a stride of 29.0 inches (74.0 cm), occurs frequently.

Scat: Seldom found because deposited below ground in burrows. Similar to, but smaller than, coyote scat, without tapered ends.

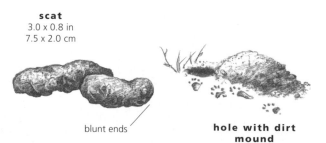

scat
3.0 x 0.8 in
7.5 x 2.0 cm

blunt ends

hole with dirt mound

SCAT WIDTH

Habitat: Open grasslands preferred. Areas with large populations of prey, which includes ground squirrels and prairie dogs.

Similar species: Differs from all other species by long claws on front foot and disproportionately small hind foot.

Other sign: Fresh excavations of large amounts of dirt from burrowing rodent holes indicates hunting activity, especially if excavated material includes large clods or rocks. Freshly widened burrow entrances may have a slightly elliptical shape. The presence of coyote and badger tracks together indicates cooperative hunting.

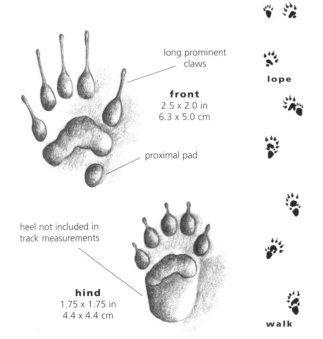

long prominent claws

front
2.5 x 2.0 in
6.3 x 5.0 cm

proximal pad

lope

heel not included in track measurements

hind
1.75 x 1.75 in
4.4 x 4.4 cm

walk

FRONT TRACK LENGTH

FRONT TRACK WIDTH

Striped Skunk
Mephitis mephitis

Black-and-white mustelid, size of a domestic cat, with triangular head. Weight varies from 4.0 to 10.0 pounds (2.0 to 5.0 kg). Male is slightly larger than female. Flat, wide, bushy tail with white hair on top. Long, curved claws for digging.

Track: Half dollar–size, with long front claws. Hind track looks like a little human footprint. Five toes, in 1-3-1 grouping. Little toe, on the inside of foot, sometimes does not register. Interdigital pad chevron-shaped. Proximal pad often shows. Hairless heel on hind foot.

Trail: Walking stride averages 12.0 inches (30.0 cm). Meanders and stops often when walking, leaving extra footprints in trail. Lope may be turned to the side or straight forward.

Scat: Cylindrical with blunt ends. Lacks the long taper and tendency to fold back on itself of other mustelid scat. May be composed entirely of insect parts.

blunt

scat
5.0 x 0.75 in
12.5 x 1.9 cm

fanged puncture

chewed eggs

Habitat: Not habitat-specific. Lives where burrows, cavities, or tunnels are present, including in and around buildings. Presence of insects and small mammals is critical to habitat selection.

Similar species: Differs from other species by having long, wide claws on the front foot. Smaller than badger, with front and hind feet similar in size.

Other sign: Smell of skunk musk identifies nests and burrows. Tears apart nests of small mammals. Bird eggs show four fang punctures around larger hole in shell.

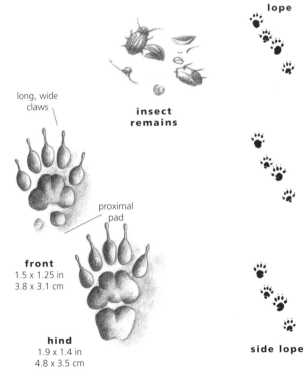

long, wide claws

insect remains

proximal pad

front
1.5 x 1.25 in
3.8 x 3.1 cm

hind
1.9 x 1.4 in
4.8 x 3.5 cm

1x2x1 lope

side lope

FRONT TRACK LENGTH

FRONT TRACK WIDTH

Snowshoe Hare
Lepus americanus

Medium-size hare with long ears and feet. Averages about 4.0 pounds (1.8 kg). Body color is rusty to gray-brown, turning white in the winter. Ears retain their black tip in winter.

Track: Toes asymmetrical around foot axis. Track indistinct because the foot is completely haired and lacks pads. Hind footprint may be up to two and a half times longer than front. To provide flotation on snow, hind feet are exceptionally wide and toes may splay apart so that width approaches length.

Trail: Hopping stride varies from 3.0 to 6.0 feet (0.9 to 1.8 m). Tends to hop with paired hind feet.

Scat: Dry scat is a slightly flattened sphere. Produces a black, semiliquid scat that is usually reingested to utilize remaining nutrients.

scat
0.3 in
0.8 cm

chewed branch and cone

SCAT WIDTH

Habitat: High mountains with deep snows. Dense second-growth forest is preferred, but swamps are also used. Forages at forest edge and in small clearings.

Similar species: Track differs from cottontail by its large, wide size.

Other sign: Look for woody plants, including conifers, that have had the tips of branches chewed off. During population highs, hares will strip tree bark and have been observed feeding on carcasses. The snowshoe's nest, a shallow depression known as a *form,* is found under conifer branches or logs.

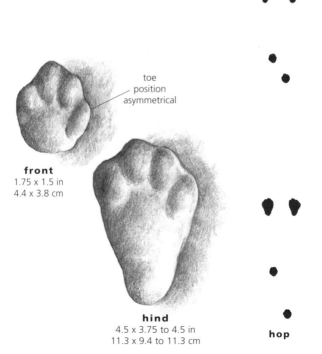

toe position asymmetrical

front
1.75 x 1.5 in
4.4 x 3.8 cm

hind
4.5 x 3.75 to 4.5 in
11.3 x 9.4 to 11.3 cm

hop

FRONT TRACK LENGTH

FRONT TRACK WIDTH

Pika
Ochotona species

Small (average weight 5.0 to
6.0 ounces/140.0 to 170.0 g),
rat-size, short-legged, nearly
tailless, egg-shaped member
of the rabbit order
(*Lagomorpha*). Large ears
and relatively large eyes.
Color gray to grayish brown.
The pika of Alaska and the
Yukon is the collared pika (*O. collaris*) while the pika found
in British Columbia is the American pika (*O. Princeps*).

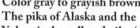

Track: Shows four toes. Only in a very clear
footprint will the minute inside toe on the
front foot be observed. Unlike the rest of
the rabbit order, the pika's track shows toe
pads. The sole is haired.

Trail: Prefers a hop, with a stride of 15.0
inches (38.0 cm). Even when moving fast,
the stride is seldom more than three times
the group.

scat
0.1 in
0.3 cm

hay pile

SCAT WIDTH

Scat: Produces two types of scat. The most readily found are nearly spherical and dry. Also produces a black, semiliquid scat that is usually reingested to utilize remaining nutrients.

Habitat: Found almost exclusively in the rock and talus fields of high mountains.

Similar species: Differs from chipmunk and other rodents by having only four toes on hind foot. Track is larger than mouse or vole.

Other sign: Marks territory by urinating on prominent rocks, leaving a white, hard stain. Scat may be placed nearby. Stores plants under and around rocks near the center of its territory. These hay piles dry and provide food for the pika during the winter.

front
0.75 x 0.6 in
1.9 x 1.5 cm

hind
1.0 x .075 in
2.5 x 1.9 cm

hop

FRONT TRACK LENGTH

FRONT TRACK WIDTH

Arctic Ground Squirrel
Spermophilus parryii

Larger than a small rat, 1.0 to 1.5 pounds (450.0 to 700.0 kg) and standing 12.0 inches (30.0 cm) tall. Males larger and both sexes put on weight as hibernation nears. White spots on a reddish-brown back but turning grayish in fall. Calls include a shrill whistle and chirp-chirp sound.

Image not available.

Track: Four toes with 1-2-1 spacing on front foot and five with 1-3-1 spacing on hind foot. Toes relatively slender. Front foot larger than a quarter. Hind heel is hairless and may register clearly in track. Claws may show.

Trail: Bounding stride about 20.0 inches (50.0 cm). Uses a half bound.

Scat: Small ovals, usually not connected.

SCAT WIDTH

scat
0.1 in
0.3 cm

**burrow
entrance**

Habitat: Arctic ground squirrel found in higher mountain meadows and tundra above treeline. Often dens in ridges around polygonal or patterned ground.

Similar species: Only ground squirrel of this northern region.

Other sign: Look for burrow entrances.

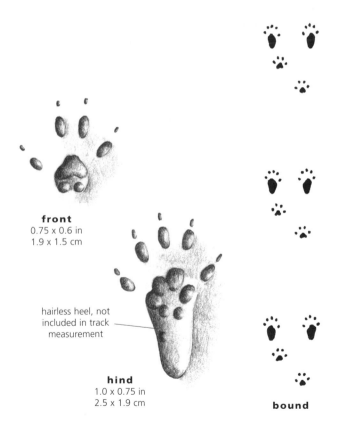

front
0.75 x 0.6 in
1.9 x 1.5 cm

hairless heel, not included in track measurement

hind
1.0 x 0.75 in
2.5 x 1.9 cm

bound

FRONT TRACK LENGTH

FRONT TRACK WIDTH

Least Chipmunk
Tamias minimus

Slightly larger than a large mouse, up to 3.0 ounces (80.0 g). Reddish fur, with white stripes bordered by black stripes along the sides of the face and body. Haired tail.

Track: Front foot size of a nickel, with four toes in 1-2-1 grouping. Five toes on hind foot, 1-3-1 grouping. Toes relatively slender. Claws short. Hind heel is haired and details are difficult to detect.

Trail: Bounding stride averages 7.0 inches (18.0 cm). Mostly terrestrial, it usually uses a half bound, though full bounds may be observed in its trails.

scat
0.1 in diameter
0.3 cm

SCAT WIDTH

Scat: Small, usually unconnected ovals.

Habitat: Deciduous forest and brush areas.

Similar species: Smaller than ground and tree squirrels. Lacks the long claws of ground squirrel. Smaller than woodchuck.

Other sign: Seeds and nuts of various plants, chewed open on one side.

front
0.5 x 0.4 in
1.3 x 1.0 cm

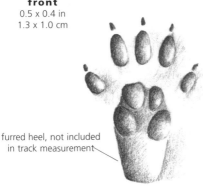

furred heel, not included in track measurement

hind
0.7 x 0.6 in
1.8 x 1.5 cm

bound

FRONT TRACK LENGTH

FRONT TRACK WIDTH

Red Squirrel
Tamiasciurus hudsonicus

Medium-size
squirrel, weighing
up to 0.5 pound
(0.25 kg).
Reddish brown
back, separated
from white underparts by a black stripe. White ring around
eye. Slight ear tufts. Tail is bushy.

Track: Front foot size of a quarter, with four toes in 1-2-1 grouping and five pads. Five toes on hind foot, 1-3-1 grouping and four pads. Toes relatively slender. Claws relatively short. Haired hind heel is indistinct in tracks.

Trail: Full bounding stride averages 24.0 inches (60.0 cm). Walk and half bound when searching for food.

Scat: Small, shapeless black masses to small, usually unconnected ovals.

scat
0.4 x 0.2 in
1.0 x 0.5 cm

**chewed cones
and seed casings**

SCAT WIDTH

Habitat: Boreal or northern coniferous forests, mixed hardwood forests, and swamps. Infrequently found in deciduous forests.

Similar species: Larger than chipmunk. Lacks the long claws of ground squirrel. Smaller than gray squirrel and woodchuck.

Other sign: Builds twig-and-leaf nests in branches of trees, about 15.0 feet (5.0 m) from the ground. Piles, called *middens,* of pine and hemlock cone scales where squirrel removes scales to get at seeds. Cones are cached 12.0 inches (30.0 cm) deep in the midden for use as winter food. Stripped buds of spruce trees scattered along forest floor.

hind
prints on
front

slow bound

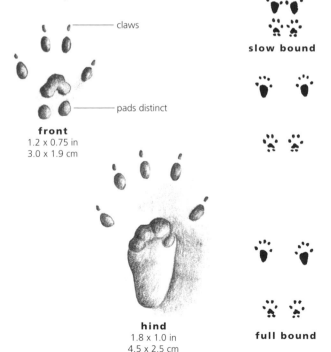

claws

pads distinct

front
1.2 x 0.75 in
3.0 x 1.9 cm

hind
1.8 x 1.0 in
4.5 x 2.5 cm

full bound

FRONT TRACK LENGTH

FRONT TRACK WIDTH

Northern Flying Squirrel
Glaucomys sabrinus

A small squirrel, weighing
4.0 ounces (112.0 g). Its
silky fur is olive brown
on the back and
lead gray on the
underside. A fold of
skin stretches between front
and hind legs and body, forming a
wing and allowing the squirrel
to glide. Its bushy tail is flattened to aid in sailing.

Track: Front foot size of a quarter, with four toes in 1-2-1 grouping. Five toes on hind foot, 1-3-1 grouping. Toes relatively slender. Claws relatively short and may not show. Hind foot interdigital pads form a tight crescent, though proximal pads are lacking.

Trail: Bounding stride averages 20.0 inches (50.0 cm). Uses a full bound.

Scat: Small, usually unconnected ovals.

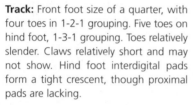

scat
0.1 in
0.3 cm

wing marks

Habitat: Deciduous and coniferous forests, though often found in attics of houses.

Similar species: Differs from all other squirrels and chipmunks by the tight crescent of interdigital pads on the hind foot. Lacks the long claws of ground squirrels. Smaller than woodchuck.

Other sign: Skin flap outlines may show in dust or snow. Sometimes builds roof on bird nest to use as den. Tree dens may hold twenty squirrels during the winter.

hind prints
on front

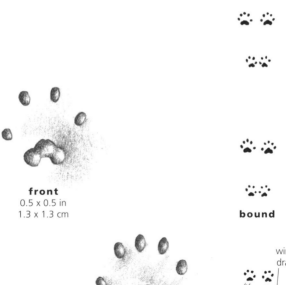

front
0.5 x 0.5 in
1.3 x 1.3 cm

bound

interdigital pads
arrayed in a
crescent shape

hind
1.5 x 0.5 in
3.8 x 1.3 cm

wing
drag

FRONT TRACK LENGTH

FRONT TRACK WIDTH

Deer Mouse
Peromyscus maniculatus

Small mouse, weighing up to 1.0 ounce (28.0 g). Adults are reddish brown to brown on back with a white belly, while juveniles are dark gray on the back with a light gray belly. Large eyes and ears. Tail is long and haired.

Track: Track smaller than a dime. Four toes on front foot, in 1-2-1 grouping. Five toes on hind foot, 1-3-1 grouping. Four joined interdigital and two proximal pads on front footprint; five joined pads and heel on hind footprint. Heel is hairless.

Trail: Bounding stride averages 8.0 inches (20.0 cm). Those species that use a full bound are climbers and nest in grass, shrubs, or trees. Those species using a half bound nest on or below ground. Both types occasionally trot. Tail drag may be present.

scat
0.08 in
0.2 cm

SCAT WIDTH

Scat: Oval-shaped pellets similar to those left by house mice.

Habitat: Ubiquitous, being found from deserts to the northern tree line, from below sea level to the top of high peaks.

Similar species: Differs from shrew by having only four toes on the front feet and by being slightly larger. Differs from vole by often showing a tail drag and by most often bounding. Lacks the long heel of the jumping mouse. Smaller than chipmunk.

Other sign: Compact grass nests without entrances may be found under logs, rocks, and boards. Enters and exits through the grass wall, which closes up after passage. Caches large quantities of seeds in any convenient protected area. Leaves feces near and in nest.

tail drag

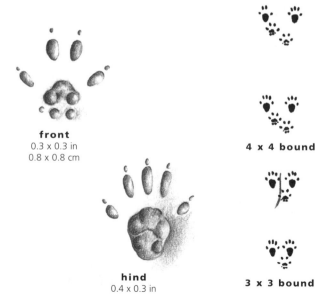

front
0.3 x 0.3 in
0.8 x 0.8 cm

4 x 4 bound

hind
0.4 x 0.3 in
1.0 x 0.8 cm

3 x 3 bound

FRONT TRACK LENGTH

FRONT TRACK WIDTH

Long-tailed vole
Microtus longicaudus

Mouse-size mammal related to the lemming and weighing up to 1.0 ounce (28.0 g). *Microtus* **are gray to gray-brown on back with a light colored belly. Stocky small mammals with short ears and small eyes almost hidden by their fur. Most voles have short, sparsely haired tails but that of the long-tailed vole is over 2.0 inches (5.0 cm).**

Track: Diminutive tracks smaller than a dime. Four toes on front foot, in 1-2-1 group. Five on hind foot, 1-3-1 grouping. Four joined interdigital and two proximal pads on front footprint and four interdigital pads and one proximal on hind footprint. Heel is hairless.

Trail: Trotting stride is 6.0 inches (15.0 cm). Usually trots and seldom bounds. Tail usually does not show in the trail.

Scat: Typical mouse-shape ovals. Often placed in tennis ball–size latrines consisting of thousands of pellets.

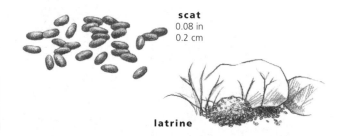

scat
0.08 in
0.2 cm

latrine

SCAT WIDTH

Habitat: Grass-loving species found near meadows and water holes in dry country.

Similar species: Differ from shrews by having only four toes on the front feet. Differ from mice by seldom showing a tail drag and usually trotting.

Other sign: In the spring as snow melts, look for grass nests lacking entrances but having cords of grass and debris. Cords were made when grass and debris were stuffed inside tunnels in the snow. Latrines are usually found near nests. Voles make worn runways through the grass.

bound

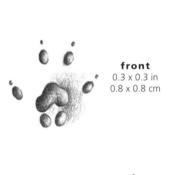

front
0.3 x 0.3 in
0.8 x 0.8 cm

fast trot

hind
0.4 x 0.3 in
1.0 x 0.8 cm

five pads
only

heel not included in
track measurement

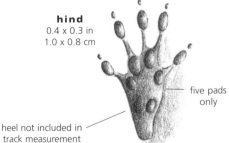

trot

FRONT TRACK LENGTH

FRONT TRACK WIDTH

Collared Lemming
Dicrostonyx groenlandicus

Large hamster-size rodents of the far north, up to 4.5 ounces (130.0 g). Another lemming, the brown lemming (*Lemmus trimucronatus*), inhabits Alaska and the Yukon. Lemming bodies are stout with blunt noses, short tails, and small ears and tails. The collared lemming, dark gray in summer, turns white in the winter and grows a digging claw.

Track: Track about the size of a dime, with hind footprint longer. Four toes on front foot, in 1-2-1 grouping. Five toes on hind foot, 1-3-1 grouping. Four joined interdigital and two proximal pads on front footprint and four interdigital pads and one proximal on hind footprint. More track and scat measurements are needed. We have only recorded blurred tracks in the snow.

Trail: Walking stride is up to 4.0 inches (10.0 cm). Straddle is 1.1 inches (2.8 cm). Brown lemmings usually walk or trot, collared lemmings often lope. Tails usually do not show in the trail.

scat
0.1 in
0.3 cm

SCAT WIDTH

Scat: Oval-shaped pellets similar to those left by house mice, often piled in tennis ball–size latrines that may hold hundreds of pellets.

Habitat: High dry areas with rocks and grass. Brown lemmings are found in tundra and alpine meadows.

Similar species: Differ from voles by having larger tracks and wider straddles on the trails. Differ from shrews by having only four toes on the front feet. Differ from mice by seldom showing a tail drag and by most often walking. Lack the long heel of the jumping mouse.

Other sign: Piles of scat are common and usually mark winter latrines.

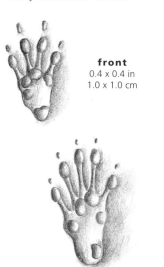

front
0.4 x 0.4 in
1.0 x 1.0 cm

hind
about 0.6 x 0.4 in
about 1.5 x 1.0 cm

walk

FRONT TRACK LENGTH

FRONT TRACK WIDTH

Western Jumping Mouse
Zapus princeps

A small—about 0.8 ounce
(25.0 g)—mouse with long
hind feet and long, sparsely
haired tail. Yellowish sides,
darker brown back,
and white belly.
White tip on long
tail.

Track: Four toes on front foot, in 1-2-1 grouping. Hind foot is about the size of a quarter, exceptionally long and narrow, and has five toes in 1-3-1 grouping. Toes relatively slender. Heel is hairless.

Trail: Bounding stride 60.0 to 120.0 inches (150.0 to 300.0 cm). Makes sharp turns during travel. May cover considerable distance per stride when pursued. Tail drag often observed.

scat
0.08 in
0.2 cm

SCAT WIDTH

Scat: Small oval pellets.

Habitat: Mountains, seldom found more than 3.0 feet (1.0 m) from a stream.

Similar species: Differs from other rodents in having long, narrow hind feet and tail drag. Differs from kangaroo rat by bounding from all four feet, not just hind.

Other sign: Small piles of grass stems left after eating. Round grass nests.

tail drag

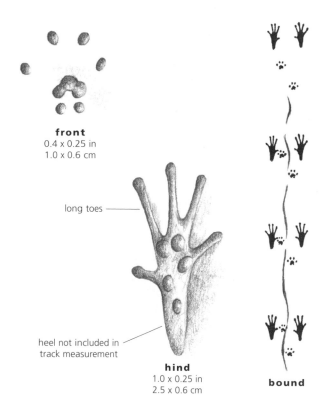

front
0.4 x 0.25 in
1.0 x 0.6 cm

long toes

heel not included in track measurement

hind
1.0 x 0.25 in
2.5 x 0.6 cm

bound

FRONT TRACK LENGTH

FRONT TRACK WIDTH

Northern Pocket Gopher
Tomomys talpoides

Guinea pig–size rodent with minute eyes and ears and a short tail. Has external, fur-lined cheek pouches.
Averages about 12.0 ounces (340.0 g), with males larger than females. Yellowish brown to dark, almost black. Two distinct grooves down front teeth.

Track: Five toes front and hind foot. Toes are relatively slender. Front feet have relatively long, wide claws for digging. Claw length is equal to or longer than toe length. Good tracks are seldom found and more data are needed on track size. Measurements are approximate.

Trail: Walking stride is 6.0 inches (15.0 cm).

Scat: Thick, short cords.

Habitat: Needs deep sandy soils preferably associated with grasslands, meadows, and fields.

scat
0.2 in
0.5 cm

SCAT WIDTH

Similar species: Differs from other rodents by large, wide claws.

Other sign: Summer mounds consist of loose dirt forming a flat mound with no entrance visible (gophers close the tunnel as they go back underground). Winter casts of soil and rocks show where gophers packed dirt into snow tunnels while they burrowed for food. Scats are often found in the tunnel casts.

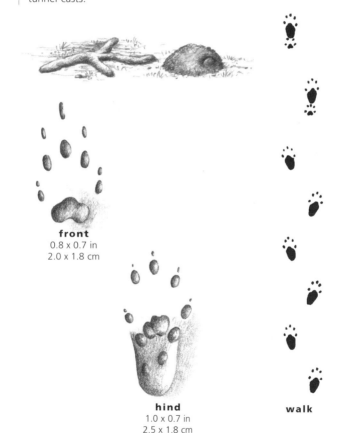

front
0.8 x 0.7 in
2.0 x 1.8 cm

hind
1.0 x 0.7 in
2.5 x 1.8 cm

walk

FRONT TRACK LENGTH

FRONT TRACK WIDTH

Muskrat
Ondatra zibethica

Large, ratlike, stocky, up to 4.0 pounds (2.0 kg). Males slightly larger than females. Small eyes and ears. Tail is black, flattened, scaly, with few hairs.

Track: Four toes on front foot (small fifth nubbin may show in very clear tracks) and five on hind foot. Toes very slender. Hind foot appears wider than long.

Trail: Walking strides average 11.0 inches (28.0 cm). May lope with body turned to side.

Scat: Oval, at most three to four times longer than wide. Often deposited in a sticky mass on exposed logs at water's edge.

Habitat: Marshes and lake edges, secondarily on streambanks. Large rivers are not as frequently used. Cattails and rushes predominate.

scat
0.2 in
0.5 cm

SCAT WIDTH

Similar species: Differs from beaver by smaller size and lack of webbing. Differs from mink by long slender toes and by usually walking.

Other sign: Small conical domes made from reeds serve as dens. Cut grass and reeds near water's edge mark feeding sites. Muskrats make "post offices," repeated scat deposits, on rocks.

4 x 4 bound

3 x 3 bound

post office

side lope

front
1.3 x 1.2 in
3.3 x 3.0 cm

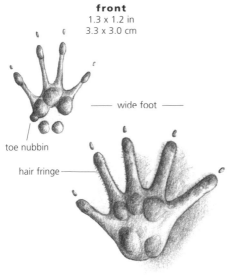

—— wide foot ——

toe nubbin

hair fringe

fast walk

hind
1.3 x 1.6 in
3.3 x 4.0 cm

walk

FRONT TRACK LENGTH

FRONT TRACK WIDTH

Beaver
Castor canadensis

Largest rodent in
North America,
30.0 to 60.0 pounds
(14.0 to 27.0 kg).
Distinguished by large,
webbed hind feet and large,
horizontally flattened tail. Fur overall is dark brown to
almost black, with lighter belly.

Track: Front and hind prints show five toes. Hind foot may be larger than a human hand. Webbing between hind toes shows, but only when pulled tight by splaying of toes. Clear tracks are difficult to find, as the hind foot steps on the front foot and the dragging tail obliterates many prints.

Trail: Walking stride 18.0 inches (45.0 cm).

Scat: Seldom found, as they are usually deposited in water, where they disintegrate quickly. Marshmallow-size, a little longer than thick. Consist of wood chips.

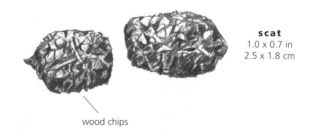

scat
1.0 x 0.7 in
2.5 x 1.8 cm

wood chips

Habitat: Seldom found far from a creek, river, pond, or lake.

Similar species: Differs from other rodents by large size and webbing. Differs from river otter by long, slender toes and pointed heel, and by lacking a chevron-shaped pad.

Other sign: Dams and conical lodges, built of twigs and sticks. Standing, cut-off tree trunks end in a tapered cone. Debarked tree limbs in the water.

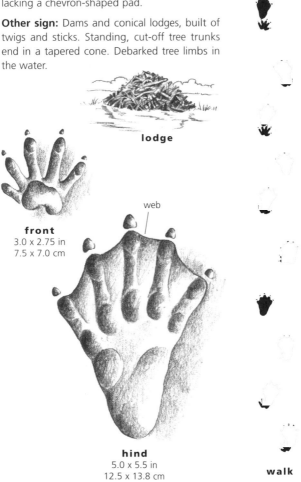

lodge

front
3.0 x 2.75 in
7.5 x 7.0 cm

web

hind
5.0 x 5.5 in
12.5 x 13.8 cm

walk

Porcupine
Erethizon dorsatum

Basketball-size or larger, 10.0 to 25.0 pounds (5.0 to 11.0 kg). Stocky body, with short legs. Distinguished by the presence of quills. Brown to yellowish brown in color.

Track: Rough texture formed by small nubs on soles of feet. Four toes on front foot and five toes on hind. Toes often do not show. Claws often show.

Trail: Walking stride 17.0 inches (43.0 cm). Tail drag often present.

Scat: Winter scat formed from feeding on conifers is red. Summer scat includes

scat
0.5 in
1.3 cm

debarked stick with chew marks

more herbs and shrubs and is brown to black. Scat from both seasons may be composed of individual pellets or strings of pellets connected by fibers.

Habitat: Generally found near forests, but may be far from trees if shrubs are available.

Similar species: Rough texture on sole of foot is diagnostic. In snow, trough made by dragging belly highlights its stockiness, separating it from faster-moving mammals.

Other sign: Twigs with bark chewed off, found at the bases of trees. Will perch in a tree for days, chewing the bark, thereby killing the tree.

tail drag

amble

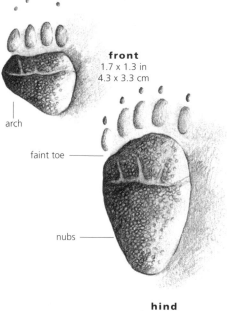

arch

front
1.7 x 1.3 in
4.3 x 3.3 cm

faint toe

nubs

hind
2.7 x 1.7 in
6.8 x 4.3 cm

walk

FRONT TRACK LENGTH

FRONT TRACK WIDTH

Moose
Alces alces

Largest member of
deer family.
Considerable regional
variation exists, but
male may weigh 900.0
pounds (400.0 kg), with
female smaller. Color
varies from tan to blackish.
Females have a white patch
of hair around the vulva (visible at a
distance) that helps identify sex. Male
has antlers that are shed annually.

Track: Long, with the pad extending to
near the front of the hoof. Subunguinis
region is narrow. Track is delicate for the
weight of the animal.

Trail: Walking stride is 70.0 inches
(175.0 cm). Seldom gallops. Trots when
in a hurry.

Scat: Most of the year, consists of dry
pellets that scatter on impact with the ground. When the diet is
moist, nipple-dimple shape predominates. Winter scat is oval, and
consists mostly of chips of woody vegetation.

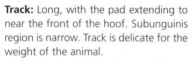

scat
pellet 0.6 in
pellet 1.5 cm

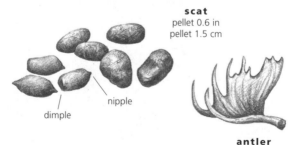

nipple

dimple

antler

SCAT WIDTH

Habitat: Mixed conifer and hardwood forests containing willows and aspen or poplar. Streams and shallow lakes provide aquatic vegetation.

Similar species: Differs from elk in that the pad occupies most of each clout. Differs from deer by being larger and more pointed.

Other sign: In removing velvet from their antlers, bulls strip bark from young saplings and break off limbs, often killing the trees. Height of tree wound shows animal height.

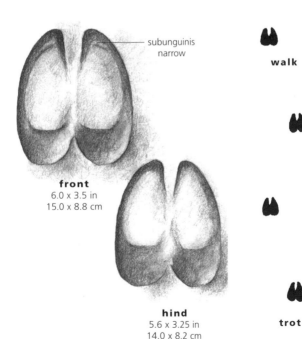

subunguinis narrow

front
6.0 x 3.5 in
15.0 x 8.8 cm

hind
5.6 x 3.25 in
14.0 x 8.2 cm

walk

trot

FRONT TRACK LENGTH

FRONT TRACK WIDTH

Caribou
Rangifer tarandus

Smaller than elk, larger than deer, with males averaging 250.0 pounds (115.0 kg) and females about 225.0 pounds (100.0 kg). The tines (points) over the nose are palmate (webbed). Fur color is gray to grayish brown. Males and females have antlers that are shed annually. Those of males are larger.

Track: Track is wide or wider than long. Hoof walls rounded on the outside, making hoof appear almost circular. Dewclaw impressions are widely spaced.

Trail: Walking stride is 48 inches (120 cm). Trotting stride 75 inches (188 cm). Caribou walk when grazing, but when moving longer distances, they tend to trot, with hind foot overstepping the front.

Scat: Typical scat is dry, falling apart when it hits the ground.

scat
pellet 0.4 in
pellet 1.0 cm

antler

SCAT WIDTH

Habitat: The southern limit of range barely reaches the lower 48 states. Habitat includes mountain summits above tree line and alpine meadows interspersed with open subalpine forests.

Similar species: Caribou track differs from other hoofed animals by its round, wide print and widely spaced dewclaws. Rounded tips of hoof distinguish caribou prints from those of bison.

Other sign: Large herds leave wide travel route. Snow may be scraped from ground to get at lichens, which caribou eat.

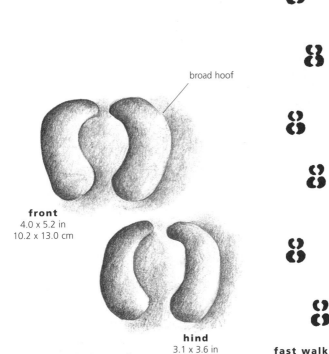

broad hoof

front
4.0 x 5.2 in
10.2 x 13.0 cm

hind
3.1 x 3.6 in
8.0 x 9.0 cm

fast walk

FRONT TRACK LENGTH

FRONT TRACK WIDTH (50%)

Mule Deer
Odocoileus hemionus

Small member of the deer
family, male averaging 160.0
pounds (70.0 kg) and female
about 130.0 pounds (60.0 kg).
Coat color is reddish brown in
summer and grayish brown in
the winter. Antlers,
found only on males,
branch symmetrically and are
shed annually.

Track: Heart-shaped, with convex wall. Pad occupies most of the clout; subunguinis slender.

Trail: Walking stride is 36.0 inches (90.0 cm). Pronks or stots with front and hind feet striking the ground at the same time. Gallops when in a hurry.

Scat: Usually dry, falls apart when lands on the ground. Pellets vary from nipple-dimple shape to oval.

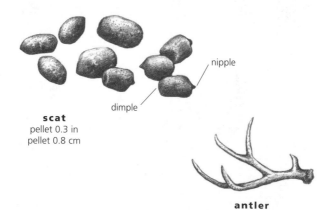

nipple

dimple

scat
pellet 0.3 in
pellet 0.8 cm

SCAT WIDTH

antler

Habitat: Foothills are prime habitat, where can frequent open brush interspersed with rugged terrain. Found in all vegetation zones except in the Arctic and in extreme desert.

Similar species: Track is not distinguishable from white-tailed deer, although mule deer is usually larger. Differs from pronghorn, goats, and sheep by having convex walls. Smaller than elk, has more slender tips, and pad occupies most of clout.

Other sign: Breaks off limbs of trees when removing the velvet from antlers. Velvet is difficult to find, as both deer and rodents eat the nutrient-rich material. Height of tree wound indicates height of animal.

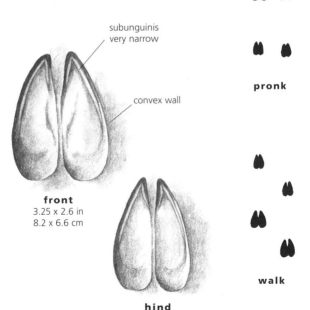

subunguinis
very narrow

convex wall

front
3.25 x 2.6 in
8.2 x 6.6 cm

hind
3.1 x 2.5 in
7.8 x 6.3 cm

pronk

walk

FRONT TRACK LENGTH

FRONT TRACK WIDTH

Elk
Cervus elaphus

Medium-size, larger than deer, males averaging 700.0 pounds (315.0 kg) and females 450.0 pounds (200.0 kg). Reddish to dark brown, with a yellow rump patch. Males shed antlers annually. Also known as wapiti. Reintroduced in east.

Track: Blocky, with each clout wide at the leading tip. Pad of hoof occupies rear third of each clout; subunguinis occupies remaining two-thirds of each clout.

Trail: Walking stride 52.0 inches (130.0 cm). When chased by a predator, gallops and occasionally pronks.

Scat: Most of the year, scat consists of pellets that scatter on impact with the ground. When the diet is moist, nipple-dimple shape predominates, changing to oval as vegetation dries. When scat is moist, pellets stick together.

nipple

dimple

scat
pellet 0.5 in
pellet 1.3 cm

antler

SCAT WIDTH

Habitat: Forest. Beds in dense trees during the day, moving out into clearings to graze during twilight hours.

Similar species: Differs from deer and moose by having a small pad at the rear of the hoof.

Other sign: Removing antler velvet, bulls strip bark from young saplings and break off limbs, often killing the trees. Height of tree wound shows animal height. Bulls make mud wallows in the fall. During rut, look for areas where bulls have sparred with the ground using their antlers.

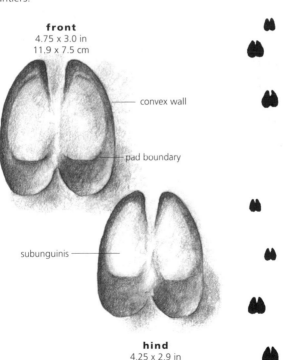

front
4.75 x 3.0 in
11.9 x 7.5 cm

convex wall

pad boundary

subunguinis

hind
4.25 x 2.9 in
10.6 x 7.3 cm

pronk

gallop

FRONT TRACK LENGTH

FRONT TRACK WIDTH

Dall Sheep
Ovis dalli

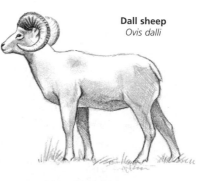

Dall sheep
Ovis dalli

Medium-size sheep, distinguished by massive horns of the male. Male averages 180.0 pounds (80 kg), females are about two-thirds the size of the male. Two subspecies exist: *O. d. dalli* is uniformly white and *O. d. stonei* is grayish brown with a white muzzle.

Northern populations consist of white Dall sheep, and southern populations consist of Stone sheep. Stone sheep are slightly larger than Dall sheep. *O. dalli* are thinhorn sheep; bighorn sheep exist in southern British Columbia. Ram has spiraled horn; ewe's is small and straight. Horns are not forked and are never shed.

Track: Blocky, with edges of the walls straight along the sides.

Trail: Walking stride is 35.0 inches (90.0 cm). Stride of a galloping Dall sheep, perhaps pursued by a wolf, was 106.0 inches (270.0 cm).

Scat: More likely to be dry and to separate

SCAT WIDTH

scat
0.3 in
0.8 cm

into pellets than that of other hoofed mammals.

Habitat: High mountain areas, especially along cliffs. Comes down from cliffs for water, and grazes on grass in rolling hills. Migrates to lower range during the winter.

Similar species: Wall differs from pronghorn and deer by having a straight edge. Shows dewclaws, which pronghorn do not. Differs from deer and elk by relatively large subunguinis region of the clout.

Other sign: Beds scraped in soil along the edges of cliffs; deposits of old scat at repeatedly used beds may be considerable. Mineral and salt licks often serve as a focus of activity.

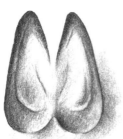

front
about 5.9 x 3.3 in
14.8 x 8.3 cm

trot

Muskox
Ovibos moschatus

Large bisonlike mammal weighing 400.0 to 900.0 pounds (180.0 to 400.0 kg). Cashmerelike brown wool hangs nearly to the ground. Light gray saddle on shoulders. Few people ever see this creature of the north; my experience with them was in Greenland.

Track: Each half of the hoof is rounded, with the over impression being of a rounded footprint.

Trail: Walking stride is about 40.0 inches (100.0 cm).

Scat: Dry, large pellets and cowpielike in early summer.

Habitat: Grassy areas in summer but seeks out wind-clear spots in the winter.

scat
pellet 0.5 in
pellet 1.3 cm

SCAT WIDTH

Similar species: Hoof is more robust than that of the caribou.

Other sign: Main indicators of muskoxen are large deposits of scat and wallowlike beds.

front
about 5.5 x 5.0 in
13.8 x 12.5 cm

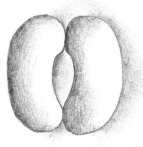

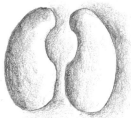

hind
about 4.5 x 4.0 in
11.3 x 10.0 cm

walk

FRONT TRACK LENGTH

FRONT TRACK WIDTH (50%)

Mountain Goat
Oreamnos americanus

Sheep-size mammal, found only in North America. Male averages 300.0 pounds (135.0 kg), female about 200.0 pounds (90.0 kg). Predominantly white, and distinguished by a stocky build and hump on shoulders. Horns, which are not shed, are straight, about 10.0 inches (25.0 cm). Those of female slightly smaller.

Track: Footprint is blocky. Tip of the clout occurs in the middle of each clout, not to the inside as in other hoofed mammals. Relatively large subunguinus.

Trail: Walking stride is about 30.0 inches (75.0 cm). Goats most often walk.

Scat: Tends to be dry and separate when it hits the ground.

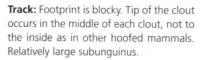

scat
pellet 0.4 in
pellet 1.0 cm

SCAT WIDTH

horn

Habitat: High mountains, on the steepest crags and cliffs. May bed among rocks, or in snow-banks or vegetated areas. Caves may be used as shelter from sun or wind. Windblown slopes are used for feeding in the winter.

Similar species: Differs from all other hoofed mammals by the location of tip in the middle of each clout.

Other sign: Digs dry wallows (shallow depressions) in the summer. Mineral or salt licks attract activity.

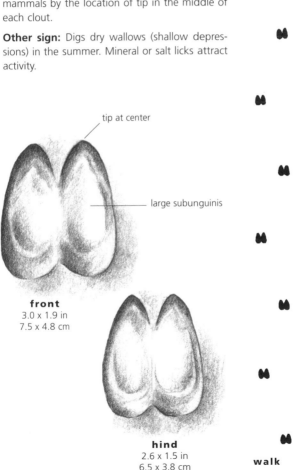

tip at center

large subunguinis

front
3.0 x 1.9 in
7.5 x 4.8 cm

hind
2.6 x 1.5 in
6.5 x 3.8 cm

walk

FRONT TRACK LENGTH

FRONT TRACK WIDTH

Selected reading

Tracks and tracking

Bang, P. et al. 1972. *Collins Guide to Animal Tracks and Signs.* London: Collins Sons.

Brown, R., J. Ferguson, M. Lawrence, and D. Lees. 1987. *Tracks and Signs of the Birds of Britain and Europe: An Identification Guide.* Kent, England: Christopher Helm.

Elbroch, M. 2003. *Mammal Tracks and Sign: A Guide to North American Species.* Mechanicsburg, PA: Stackpole Books.

—. 2001. *Bird Tracks and Sign: A Guide to North American Species.* Mechanicsburg, PA: Stackpole Books.

Fjelline, D. P., and T. M. Mansfield. 1989. Method to standardize the procedure for measuring mountain lion tracks. In *Proceedings of the Third Mountain Lion Workshop,* ed. R. H. Smith, 49–51. Prescott, AZ: Arizona Game and Fish Department.

Forrest, L. R. 1988. *Field Guide to Tracking Animals in Snow.* Harrisburg, PA: Stackpole Books.

Halfpenny, J. C. 1997. *Tracking: Mastering the Basics.* 180 min. A Naturalist's World. Videocassette.

—. 1986a. *A Field Guide to Mammal Tracking in North America.* Boulder, CO: Johnson Books.

—. 1986b. Tracks and Tracking: A "How to" Guide. Gardiner, MT: A Naturalist's World. Slides.

Halfpenny, J. C. et al. 1996. Snow tracking. In *American Marten, Fisher, Lynx, and Wolverines: Survey Methods for Their Detection,* eds. W. Zielinski and T. Kucera, 91–163. General Technical Report PSW-GTR-157. Berkeley, CA: USDA Forest Service, Pacific Southwest Research Station.

Headstrom, R. 1971. *Identifying Animal Tracks: Mammals, Birds, and Other Animals of the Eastern United States.* New York: Dover.

Lowery, J. C. 2006. *The Tracker's Field Guide.* FalconGuide. Guilford, CT: Globe Pequot Press.

Murie, O. 1954. *A Field Guide to Animal Tracks.* Peterson Field Guide Series, no. 9. Boston: Houghton Mifflin.

Rezendes, P. 1999. *Tracking and the Art of Seeing: How to Read Animal Tracks and Sign.* 2nd ed. Charlotte, VT: Camden House.

Seton, E. T. 1958. *Animal Tracks and Hunter Signs.* New York: Doubleday.

Recommended field identification guides

Burt, W. H., and R. P. Grossenheider. 1964. *A Field Guide to the Mammals.* Peterson Field Guide Series no. 5. Boston: Houghton Mifflin.

Kays, R. W., and D. E. Wilson. 2002 *Mammals of North America.* Princeton, NJ: Princeton University Press.

National Geographic Society. 1983. *Field Guide to the Birds of North America.* Washington, D.C.: National Geographic Society.

Peterson, R. T. 1984. *Birds of the Eastern United States.* Norwalk, CT: The Easton Press.

Sibley, D. A. 2003. *The Sibley Field Guide to Birds of Western North America.* New York: Alfred A. Knopf, Inc.

Stebbins, R. C. 1985. *A Field Guide to Western Reptiles and Amphibians.* Peterson Field Guide Series, no. 16. Boston: Houghton Mifflin.

Index

About the author

James Halfpenny has searched for dinosaur tracks in Colorado and Montana, tracked wildlife in Tanzania and Kenya, studied endangered species on China's Tibet-Qinghai plateau, and researched the polar bears of Hudson Bay and Greenland. Jim specializes in educational, environmentally oriented programs about mammal behavior, tracking, and cold. Since 1961 he has taught outdoor and environmental education for a vast array of schools and organizations, including the Smithsonian, National Outdoor Leadership School, Outward Bound, the Appalachian Mountain Club, The Wilderness Society, the National Wildlife Federation, Defenders of Wildlife, and the National Audubon Society. He has trained rangers in tracking techniques at Yellowstone, Glacier, Grand Teton, and Rocky Mountain National Parks. His research has also taken him to Antarctica, and Japan, and all over North America. He is a fellow of the Explorer's Club and a Vietnam veteran. Halfpenny was featured, with Australian aborigines, Kalahari Bushmen, and Alaskan Inuit, in a documentary about the loss of native tracking skills shown on the Discovery Channel. He is a past field director and project coordinator for the University of Colorado's Institute of Arctic and Alpine Research. He is also the senior author of the the *Scats and Tracks* series (FalconGuide), *A Field Guide to Mammal Tracking in North America* (Johnson Books), *Yellowstone Wolves in the Wild* (Riverbend Press), *Discovering Yellowstone Wolves: Watcher's Guide* (A Naturalist's World), *Winter: an Ecological Handbook* (Johnson Books), and *Bare Feet, Wet Socks: Writing about Winter* (edited with Ann Zwinger, Caldera Press). He lives just outside Yellowstone National Park in Gardiner, Montana.

About the illustrator

Todd Telander is a freelance natural
science illustrator and wildlife artist. He
studied biology and environmental studies
at the University of California, Santa Cruz,
where he became interested in illustration. His work
appears in Falcon's *America's 100 Most Wanted Birds*, *Birder's
Dictionary*, *A Field Guide to Cows*, and *A Field Guide to Pigs*, as
well as in museums, galleries, and private collections. Todd lives
in Walla Walla, Washington, with his wife, Kirsten, a writer, and
their two sons, Miles and Oliver.